THE PHYSICAL
ENVIRONMENT

World from 20,000 miles
[Applied Technology Satellite (Goddard Space Centre) N.A.S.A. (Nov. 18, 1967)].

THE PHYSICAL ENVIRONMENT

by

B.K. RIDLEY

Department of Physics, University of Essex, Colchester

ELLIS HORWOOD LIMITED
Publishers   Chichester

Halsted Press: a division of
JOHN WILEY & SONS
New York - Chichester - Brisbane - Toronto

First published in 1979 by
ELLIS HORWOOD LIMITED
Market Cross House, Cooper Street, Chichester, West Sussex PO19 1EB, England
The publisher's colophon is reproduced from James Gillison's drawing of the
ancient Market Cross, Chichester

Distributors:

Australia, New Zealand, South-east Asia:
Jacaranda-Wiley Ltd., Jacaranda Press,
JOHN WILEY & SONS INC.,
G.P.O. Box 859, Brisbane, Queensland 40001, Australia.

Canada:
JOHN WILEY & SONS CANADA LIMITED
22 Worcester Road, Rexdale, Ontario, Canada.

Europe, Africa:
JOHN WILEY & SONS LIMITED
Baffins Lane, Chichester, West Sussex, England.

North and South America and the rest of the world:
HALSTED PRESS, a division of
JOHN WILEY & SONS
605 Third Avenue, New York, N.Y. 10016, U.S.A.

© 1979 B.K. Ridley/Ellis Horwood Ltd.

British Library Cataloguing in Publication Data
Ridley, Brian K.
The physical environment. –
(Ellis Horwood series in environmental sciences).
1. Earth   2. Astronomy
I. Title
550   QE501   78-41223
ISBN 0-85312-142-7 (Ellis Horwood Ltd., Publishers)
ISBN 0470-26745-3 (Halsted Press)

Typeset in Press Roman by Ellis Horwood Ltd.
Printed in Great Britain by Cox & Wyman Ltd., London, Fakenham and Reading

To

AARON AND MELISSA

Behold the world, now it is whirlèd round!
And for it is so whirl'd, is named so;
In whose large volume many rules are found
Of this new art, which it doth fairly show.
For your quick eyes, in wandering to and fro
From east to west, on no one thing can glance
But, if you mark it well, it seems to dance.

Sir John Davies (1569–1626) *Orchestra*

Table of Contents

List of Tables

Preface

The environment, like the world, may be too much with us, but it is no longer something we can take for granted. Man's power to manipulate Nature is growing ever stronger and it is reaching out to manipulate and change more and more of the world around us. Because of that power the environment has become a subject of political activity as it has never been before, yet there is very little general awareness of its basic fabric. The enormous complexity of the activity about us, and our ignorance of detailed causes, has made it difficult to abstract the foundations from the many rooms and floors above. There is the Geography Room, the Geology Room, the Ecology Room, and modern additions devoted to Pollution and Energy Reserves; there is Meteorology, Oceanography, and even Astronomy; there are many cubicles as well. Such are the specialities, and indeed they are essential knowledge-producers, but they cannot separately provide a base on which to build a general awareness of the environment. Underlying and supporting all environmental processes is the fundamental physical structure of matter and energy, and it is this which provides an absolute framework against which to pin the rich, bewildering interplay of chance and necessity that is our environment. The main motivation for writing this book was to attempt to extract from many fields the basic physical structure, and to present it as a unity.

I hope the book will appeal to students of science and engineering, whether physical, biological or social. In order to make that appeal as general as possible I have not included any mathematics in the text. Instead, those quantitative topics which can be illuminated with mathematics (no more advanced than simple differential and integral calculus) have been displayed as mathematical diagrams, and they can be ignored entirely by the non-mathematical reader. The book is a much expanded version of a physics course given by the author to first-year science and engineering undergraduates at the University of Essex. Such pedagogical origins account for the inclusion at the back of the book of a set of problems, meant to extend the quantitative awareness of the reader rather than to assess his intellectual prowess and to commend the book for course-use in the world of teaching and learning.

I must confess that in writing this book I have been more student than savant. My own speciality in physics does not lie in one of the environmental fields and, although that could conceivably be regarded as an advantage in the present context, it has meant running obvious risks. Those risks have been mitigated by the good fortune of having Professor Richard Scorer as a splendidly enthusiastic editor and I am very grateful to him for his valuable criticism and his constructive suggestions. If there are errors still, or omissions they are mine only. I am also grateful to my wife for her remarkable good-nature in putting up with the requests for criticism of English, style and content, and her efforts in preparing a readable version of the manuscript.

B. K. RIDLEY
Colchester, 1978

1

Introduction

I must therefore give an account of celestial phenomena, explaining the movements of sun and moon and also the forces that determine events on earth.

Lucretius (*c*100–55 B.C.)
On the Nature of the Universe.

We live on a vast, rotating chunk of rock which orbits an enormous fireball of burning hydrogen we call the sun. The sun is moving at high speed towards another burning fireball that we call a star, Vega by name, and we have no choice but to tag along with it. (Figure 1.1). The sun and Vega are but two members of the 'local' groups of stars which form the constellations of the night sky. They lie in the rural suburbs of the galaxy, and partake of the stately rotation of the spiral of stars which appear to us as the Milky Way. (Figure 1.2). The galaxy itself moves on its own private odyssey, within the local cluster of galaxies, whose centre lies towards the Andromeda nebula, a telescopically spectacular galaxy of stars like our own. (Figure 1.3). All movement takes place in what is virtually a vacuum, from which we are shielded by the relatively thin skin of atmosphere glued to the earth's surface by gravity (Figure 1.4).

The whole venture is really rather hazardous. Charging involuntarily through space, with the chance of colliding with an errant member of the solar system, such as an asteroid, or with a star, or with a black hole (if such a thing exists); totally dependent upon a fireball whose minor whimsicalities could roast us or freeze us; helpless in the face of poorly predictable terrestrial convulsions; man can scarcely be regarded as a good risk in the cosmic insurance business. That the present physical environment is benign is the lucky chance that has allowed us to flourish, but we cannot count on it continuing for ever.

Indifferent, and potentially dangerous though the physical environment is to man, it is not the only source of peril. Man is only one species of life among millions, all deriving from a common stock based on the chemistry of four or five nucleotide bases, which make up the nucleic acids DNA and RNA, and twenty amino acids, which make up the numerous proteins. Because life on earth is founded on basically identical processes, all life forms require the same carbon-containing organic materials for their survival. The supply of these

Index Map

Figure 1.1(a) — The mid-
night sky at London, looking south, June 15. The
solar system is travelling
towards Vega. The centre
of the Galaxy lies in the
direction of the constel-
lation Sagittarius.
(From The Midnight Sky
by Sir Edwin Dunkin)

Index Map

Figure 1.1(b) – Looking south at midnight, May 15, at Sydney, New South Wales. The centre of the Galaxy is in Sagittarius (top left in the Milky Way).
(From the Midnight Sky by Sir Edwin Dunkin)

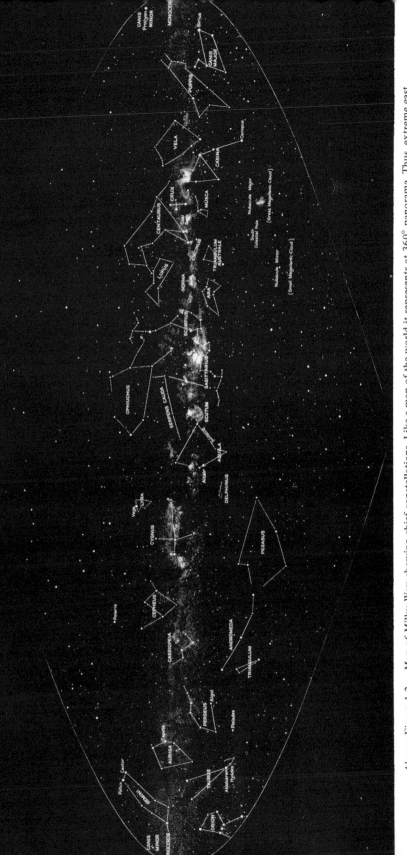

Above: Figure 1.2 — Map of Milky Way showing chief constellations. Like a map of the world it represents at 360° panorama. Thus, extreme east and west constellations appear in reality close together and in the direction opposite to that towards Scutum.

(Crown Copyright, Science Museum, London)

Below: Figure 1.3 — A nearby galaxy. The Andromeda Nebula (M.31).

(Ritchey and Pease, 1901. Photo Science Museum, London)

materials is limited and competition for them inevitable. There are no species based upon silicon, an element with chemical properties very similar to carbon, very abundant on the earth, and the closest rival to carbon in its ability to produce complex molecules. Competition between carbon-based and silicon-based organisms would not be as severe as it is in the internecine struggle within the carbon-life family. Although many mutually advantageous alliances exist between organisms and man, the biological environment of man contains a continual threat of annihilation by hostile predators, particularly in the form of viruses and bacteria, forms forever mutating. Dangers just as deadly lie in the sub-set of the biological environment formed by human society. Man has essentially the emotional equipment of the paleolithic hunter, and the technological knowledge of massive self-destruction which manifests itself in chemical, biological and nuclear weapons. He is a creature prone to rage, tribal superstitions, impressively conflicting rationalisation, and at the same time he is a dispassionately rational being, patiently acquiring insights and understanding of himself and his environment through his arts and science. But there is very little of the latter and a great deal of the former. The schizoid condition is not a symmetrical one. With the hunter dominating the sage, survival in the local biological environment is likely to depend on how fast the hunter can evolve into a less aggressive animal.

Fortunately, we have no yard-stick to measure the probability of man's surviving for, say, a further million years. If we had, we might be too much daunted by the odds against, or too much encouraged by the odds for. Our civilisation, for better or worse, is the only one we know, and in any case it may be unique even among civilisations of the universe. We are on our own, and face serious problems of survival; some chronic, some acute, some, no doubt, not even appreciated yet. Though frightfully exposed and weak, we are not entirely helpless. We have a vast energy and zest for life, and we have that intellectual compound of imagination and rationality that manifests itself in social organisation, the arts and the sciences. The full co-ordination of human energy and intellect has potentialities we can scarcely grasp. There have been hints of its power in the growth of empires and religions, in global exploration, in industrial expansion, in a nation at war and in the exploration of the moon. Only a tiny fraction of mankind has ever been involved in any one of these situations. There is an increasing awareness of this latent power which resides in mankind and, just as important, an increasing awareness of problems which affect all of mankind, and not just a special sub-set. There is also the awareness that the solution of these problems demands the coherent action of the whole human species. This in turn demands the neutralisation of the devisive elements in tribalism, racialism and political and religious creeds — the sort of neutralisation which happens frequently in the global activities of commerce, art and science. But such a mobilisation of human power, without a thorough appreciation of the subtle interconnections which exist in our environment, can be downright

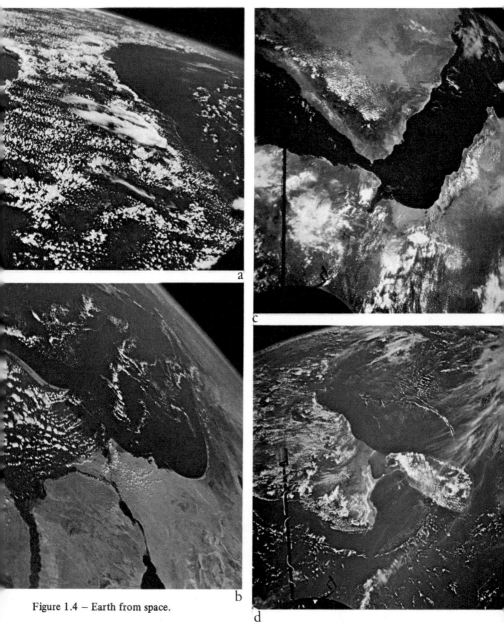

Figure 1.4 – Earth from space.

(a) Florida.   (Photo by Gemini V, August 22, 1965, NASA)
(b) Nile delta, Sinai, Israel, Syria, Jordan, Lebanon, Turkey, Cyprus.
                                                (Photo by Gemini VII, Dec. 8, 1965, NASA)
(c) Ethiopia, Somali, Saudi Arabia, Yemen, Red Sea, Gulf of Aden.
                                                (Photo by Gemini XI, Sept. 14, 1966, NASA)
(d) India, Sri Lanka, Bay of Bengal.   (Photo by Gemini XI, Sept. 14, 1966, NASA)

disastrous.

Though survival is a powerful incentive for the study of our environment, it is not the only one. Nor ought it to be, even from the standpoint that takes survival as the only criterion. An activity with a single motivation runs the risk of being blinkered. The more reasons a scientist has for carrying out an investigation, the more likely he is to identify potentially important features which a singly-motivated investigator might disregard. Tightly directed work may be fine when the end product is a device of some kind (it depends very much on what sort of device), but it is too restrictive if the ultimate goal is understanding. Survival is far too awesome a motivation. One might substitute plain enjoyment, the fascination of the acquisition of insights of the way the environment has affected life on the planet, the pleasure of tracing connections between diverse phenomena. In a word, one might play. Science, art and the organisation of people are all, in some sense, the play of human animals. It is a play with high survival value in the long run, and without it life would be a grim business indeed.

The physical environment of man is, relative to the biological environment, simple. It contains the interplay of matter and natural forces in, on, and off the planet, which at base is well-described by physics. Nevertheless, the route which lies between the basic interactions of matter and the complexities of behaviour on a macroscopic scale which the environment presents, is a long one. If it is charted at all, the route is often only dimly perceived. We have a good idea of how the seasons are related to the rotating, orbiting earth, but we are distinctly vague about the cause of ice ages. We have pretty good evidence that continents drift about, but what causes that drift is not very well understood. We understand how the rotating earth bends the wind around cyclones and anticyclones, but our prediction of the weather is still a very hazardous affair. We know that the earth's magnetic field traps energetic particles from the sun and protects us from radiation damage, but the origin of that field is problematic. We can describe in broad terms the evolution of a star, but how planets come about and how many stars have planetary systems cannot be described definitively. One thing that we can say with certainty, is that our physical environment is not confined to the earth, nor even to the solar system, but that it stretches beyond the confines of the galaxy to encompass the whole universe.

2
Day and Night

The wanderer was alone as heretofore, ...
.........................with the stars
And the quick Spirit of the Universe
He held his dialogues; and they did teach
To him the magic of their mysteries;
To him the book of Night was open'd wide
And voices from the deep abyss reveal'd
A marvel and a secret – Be it so.

Lord Byron (1788–1824) *The Dream*

2.1 HISTORY

Perhaps the most striking aspect of the physical environment is the alternation of day and night. Why this occurs is one of the oldest questions in the world, and a reasonably satisfactory answer has been found only within the last fifty years. It is an answer which forges a strong bond between conditions on earth and the farthest reaches of the universe.

The first step in the explanation involves the appreciation that the earth is spherical, and floats freely without being supported by anything. The free-floating theory was advanced by one of the earliest Greek philosophers, Anaximander of the Milesian School, who flourished in the first half of the 6th century B.C.; and Pythagoras, born about 532 B.C., was probably among the first to believe that the earth was spherical. About a hundred years later, the followers of Pythagoras obtained substantial support for their spherical earth from the theories of Anaxagoras, who taught that the moon shone by reflected light, and who gave the correct account of the eclipses. The curved shadow on the moon during lunar eclipses was strong observational evidence that the world was indeed round.

Towards the end of the 5th century B.C. the Pythagoreans made another important contribution through a Theban called Philolaus. Philolaus thought that all the planets and the earth moved in circles around some central fire. This idea was remarkable in that it put the earth on the same footing with the planets, shifting it firmly from an eternally fixed position at the centre of the universe. The word planet is derived from the Greek, meaning 'wanderer'. So to imagine the earth as just one of the 'wandering stars', like the five planets known at that time, (Mercury, Venus, Mars, Jupiter and Saturn) was quite a daring idea.

The central fire, ('the house of Zeus'), about which they orbited was not the sun, but an additional entity. The sun, indeed, shone merely by reflected light, like the moon. The ultimate source of light was not directly visible because the earth always turned the same face to the central fire, just as the moon always turns the same face to the earth. There was also a 'counter-earth', a body which manifested itself in eclipses.

It would seem perverse of Philolaus to invent gratuitously a central fire, but it was not that silly from the point of view of explaining eclipses, as the following speculation perhaps shows. Although the vast majority of the eclipses of the moon can be simply explained by supposing that the earth gets directly in between sun and moon, such a theory cannot explain eclipses of the moon when both sun and moon are observed to be above the horizon (which on occasion they certainly are) without introducing the physical effect of refraction of light by the atmosphere. Since refraction was unknown, two conclusions follow: it would appear that the sun's light was powerless to prevent the eclipse and so another source of light would have to be invented; and since this source of light was forever unobservable from the Mediterranean, it must be always situated 'behind' the earth, and it followed, therefore, that the earth itself in this case, could not be the cause of the eclipse. Under these conditions it was supposed by Philolaus that the eclipse was caused by a 'counter-earth', situated at the same distance from the central fire as was the earth. Thus, from the correct observation that eclipses of the moon occur when both sun and moon are above the horizon, it seems to the present author plausible that the foregoing chain of reasoning could account for the postulation of a central fire, distinct from the sun, and of a dark, mysterious counter-earth. But perhaps the more intense motivation of Philolaus to introduce two directly unobservable heavenly bodies into his cosmology was the fact that five planets, earth, sun, moon, counter-earth and central fire add up to ten entities, and the number ten had great magical significance for the Pythagoreans!

Further advances were made by Oenipedes and Heraclides. Oenipedes, who lived somewhat later than Anaxagoras, discerned that the ecliptic, the line across the sky followed by the planets, was inclined to the earth's axis. Heraclides (about 388-315 B.C.), a contemporary of Aristotle, saw that Mercury and Venus revolve around the sun, and believed that the earth revolved around its own axis in 24 hours.

By the end of the 4th century B.C. the answer was almost there. The final step was taken by Aristarchus of Samos (about 310-230 B.C.) who advanced the theory that the earth and planets revolve around the sun in circular orbits, and that the earth revolves about its own axis in 24 hours. Sadly, all of this tremendous achievement of Greek astronomy was effectively lost for eighteen centuries, and the whole thing had to be rediscovered by Copernicus (1473-1543).

Naturally, Greek astronomy was hampered by the lack of accurate measuring instruments, so the theory of the solar system was inevitably qualitative rather than quantitative. Nevertheless, in the 3rd century B.C. Eratosthenes, the chief librarian in Alexandria, managed to measure the earth's diameter to within about 100 km, and Ptolemy, a century later, measured the distance to the moon (30.2 earth diameters) and got a figure of 29.5 earth diameters. But the discovery that the planetary orbits were elliptical rather than circular had to await the accurate observations of the Danish astronomer Tycho Brahe (1546-1601) and their interpretation by Kepler (1571-1630). The detection of planetary satellites such as the moons of Jupiter, needed the Dutch invention of the telescope and its improvement and application by Galileo (1564-1642). And then the laws of Kepler and those of Galileo were subsumed in the sublimely simple theory of gravitation advanced by Newton (1642-1727).

The planets do indeed float freely in space, as does the sun. Each planet rotates about its own axis, and at the same time moves around the sun in an elliptical orbit. No force is required to keep the planets moving because all matter possesses a mysterious quality called inertia, which tries to keep a body's motion such that the body travels in a straight line with uniform velocity. The measure of this quality is the inertial mass. A force is therefore required to make a planet travel in an ellipse instead of a straight line, and this manifests itself as an attraction between the planet and the sun. Matter attracts matter universally by a force we call gravity. Newton showed that elliptical orbits follow from the assumption that the strength of gravity between two bodies is proportional to the product of their masses, and inversely proportional to the square of the distances between them.

Besides accounting for the motion of astronomical bodies, Newton's equation applies to an intimately familiar part of our physical environment, namely, the tendency of things, including ourselves, to fall towards the centre of the earth. It is a very real physical constraint on our movement that we grow up with and have to cope with. So familiar are its effects that it is surprising how intrinsically weak gravity is. Compared with the electromagnetic forces which bind atoms and molecules together in solids and liquids, it is tiny. The electric force between an electron and a proton in a hydrogen atom is forty powers of ten stronger than the gravitational force between them. If the sun and planets were only slightly electrically charged there would be an enormous effect upon their relative motions.

The sizeable accelerations of falling bodies on earth (9.81 ms^{-2}) merely reflects the enormous mass of the earth, 5.977×10^{24} kg. Accelerations only a little less can be produced by a few kilograms of human muscle, working on relatively weak electric impulses, at the beginning of a sprint. Weak though it is, in an electrically neutral universe, gravity is the dominant force.

Newton's theory of gravitation has been immensely successful and completely adequate for all but the most esoteric problems. That there were cases where Newtonian gravitational theory was inadequate was first realised by Einstein early this century, and his new theory of gravitation published in 1915 predicts many strange new phenomena in intense fields. One of the most dramatic is the effect of gravitation on light. Not only is the path of light bent, but if the field is intense enough light becomes trapped and cannot escape. A body dense enough to produce such a field could not be seen. It would absorb all light that fell upon it and allow none to be emitted. Against a bright background of stars it would appear as a black hole. Astronomers today are searching the heavens for such a phenomena, and so far have not found one.

Einstein's theory explains all that Newton's does and goes further. Yet it is unlikely that even this theory is the last word. The present time is remarkable for the number of competing theories which have been proposed, and experimentalists will be busy for years to come in testing them. One experiment that is of immediate significance for our physical environment is the search for gravitational waves, the gravitational analogue of electromagnetic waves. Is our solar system bathed in gravitational waves emanating from distant parts of the universe? Could such waves trigger off earthquakes? Such waves are possible according to Einstein's theory and it is important to know whether they are produced anywhere. A research team in the U.S.A. headed by Weber claims to have detected gravitational waves, and several laboratories all over the world are seeking to confirm or refute this.

2.2 OLBERS' PARADOX

Black holes and gravitational waves seem a far cry from our original concern about day and night. Surely Heraclides had the answer over 2,000 years ago. The earth revolves about its own axis every 24 hours. When the bit of the earth's surface on which we stand points towards the sun it is day, and when it points away it is night. What goes on in the rest of the universe is beside the point. The alternation of day and night is a purely local phenomenon related to the rotation of the earth in the proximity of the bright sun. Rarified gravitational theories may be needed to describe the universe at large and the odd peculiarities within it, but Newtonian mechanics is really quite adequate to describe the earth's rotation, which is what we are concerned with.

It is immensely interesting that this common-sense view is almost totally invalid. Its fallacy was pointed out by the German astronomer Olbers in 1826, in an argument that became known as Olbers' Paradox. Olbers pointed out that if the stars were uniformly distributed in space and this distribution stretched out to large enough distances there would be no direction which did not point to a star. The night sky should be therefore uniformly bright with star-light. And since stars are suns the whole sky should be as bright as the surface of the sun! The rotation of the earth should be irrelevant. Both night and day should be equally bright, as bright as the surface of a star, and we should be fried to a cinder.

There are several ways to resolve this paradox. First, it is known that the stars we see belong to a huge galaxy of stars. They are by no means uniformly distributed in space. Our sun is situated in an undistinguished position, and the bulk of the galaxy appears as the slash of faintly luminous nebulosity across the sky that we call the Milky Way. The discovery of the galaxy appeared to settle the paradox. Moreover vast interstellar gas and dust clouds obscure quite a lot of the central regions of the galaxy, further limiting the amount of star-light reaching the earth.

That this was only a temporary respite became apparent with the discovery of extra-galactic nebulae. Our Milky Way galaxy turned out to be only one of countless numbers of galaxies spread uniformly out to the furthest distances probed by our giant telescopes. If we substitute galaxy for star in Olbers' argument we arrive at the same paradox.

A quantitative look at the situation will emphasize the extraordinary nature of the paradox. Let L be the average luminosity of a galaxy. The luminosity is the amount of radiant energy emitted per second. The quantity L will be of the order 10^{11} times the luminosity of an average star, since a typical galaxy will contain about 10^{11} stars. This radiant energy spreads out uniformly in all directions (on a scale in which the size and shape of a galaxy can be neglected), and the intensity falls off according to an inverse-square law. The further away a galaxy is, the less bright it appears. This is exactly compensated by the increasing number of galaxies at larger distances, so that the total intensity remains constant. Increasing dimness is made up by increasing numbers, and the total light output from galaxies a given distance away does not depend upon that distance.

If we try to escape from the paradox by noting the presence of huge amounts of obscuring dust clouds in our own galaxy, we run into difficulties. There is appreciable absorption only of light travelling in the galactic plane, but otherwise the galaxy is transparent. But even if dust clouds did absorb the cosmic glare they would soon become so hot they would radiate themselves as brightly as stars. Absorption by galactic clouds cannot resolve the paradox. The important thing here is the dependence of distance (Figure 2.1). If we add up the contribution at all distances up to the point when one galaxy begins to mask more distant galaxies we get back to Olbers' deduction that the sky should be uniformly bright, with the surface brightness equal to that of a galaxy, at least — as if the Andromeda Nebula were smeared across the whole sky.

Replacing the surface brightness of a star with the surface brightness of a galaxy, a reduction of the order of 10^{12}, can only be achieved by abandoning the idea that the universe is changeless and infinite. An infinite universe provides infinite energy, and a changeless universe allows time for light to have travelled from any distance, and furthermore, allows thermodynamic equilibrium to be established. Under such conditions galaxies would be effectively transparent and we would return to Olbers' paradox in its original form. But since galaxies are much paler objects overall than individual stars, the universe cannot be infinite and changeless.

If we accept that and work with galaxies instead of stars we still have to explain the darkness of the night sky. The observed density of galactic matter in the universe is about one galaxy in a volume of $10^{68}m^3$, on average. Taking the area of galaxy projected towards us to be roughly that of a circle of radius 10 light-years (1 light-year $= 9.46 \times 10^{15}$m), we can work out how big the universe must be in order for the whole sky to be filled with galaxies. It turns out that the universe must have a radius of about 10^{12} light-years, which is only a hundred times beyond the range of our telescopes. We cannot therefore escape the paradox by assuming that the universe is significantly smaller than that, because observation tells us it is not. Out to 10^{10} light-years or so there is no sign of any falling off in the number of galaxies, so the night sky should be much brighter than it is.

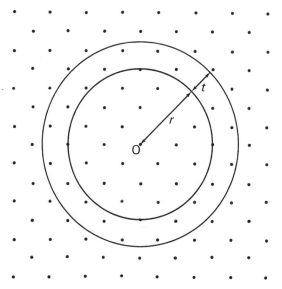

Figure 2.1a – Intensity at O from galaxies in spherical shell independent of distance.
Intensity at O from one galaxy in spherical shell $\alpha \ \dfrac{L}{r^2}$ where $L =$ luminosity.

No. of galaxies in spherical shell $= \rho V$.

Where $\rho =$ density of galaxies and $V =$ volume of spherical shell.
Now $V = 4\pi r^2 t$ if $t \ll r$.
∴Intensity at O from all galaxies in shell $\alpha \ \dfrac{L}{r^2} \times \rho \times 4\pi r^2 t$ and therefore independent of r.

Figure 2.1(b) (*following page*) – Obscuring clouds of gas are found in our own galaxy. The Horsehead Nebula south of the constellation Orion.   (Photo: Duncan, Science Museum, London)

Fortunately, there exists an observation which resolves the paradox. The observation, made some fifty years ago by Hubble, is that the spectrum of light emitted by distant galaxies is shifted towards the red (Figure 2.2). Hubble discovered that the red-shift increased in proportion to the distance of the galaxy: the further the galaxy, the greater the red-shift.

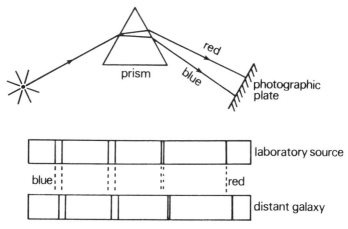

Figure 2.2 – Red shift of spectral lines.

The simplest explanation of Hubble's observation is that the red-shift is caused by the Doppler effect (Figure 2.3). Suppose we have a source of light which emits a well-defined wavelength, such as a sodium-vapour lamp, which we can measure. When the lamp moves towards us the light waves become bunched up, and we measure a shorter wavelength. When it moves away from us the waves are more extended and we measure a longer wavelength. This is the Doppler effect, and it works for sound waves as well as light waves. The pitch of a train whistle or the siren of a police car or ambulance drop noticeably as they rush past. In this case a 'blue'-shift is followed rapidly by a 'red'-shift. If we look at nearby stars, or nearby galaxies, we can often detect a spectrum shift, sometimes towards blue, sometimes towards red. If towards blue, the star or galaxy is coming towards us at a speed we can calculate from the simple equations for the Doppler effect; if towards red, the distance is increasing. But the more distant the galaxy, the more systematic is the shift towards red. The Doppler-effect interpretation means that distant galaxies are receding from us with speeds which increase with distance. The universe is expanding.

The discovery of the cosmological red-shift resolves Olbers' paradox. If the light from the most distant galaxies is largely shifted out of the visible range by recession we have an explanation of why the sky is dark at night. The answer lies, not in local conditions, but in conditions which pertain in the most distant reaches of the universe. It is dark at night because the earth is revolving and the universe is expanding!

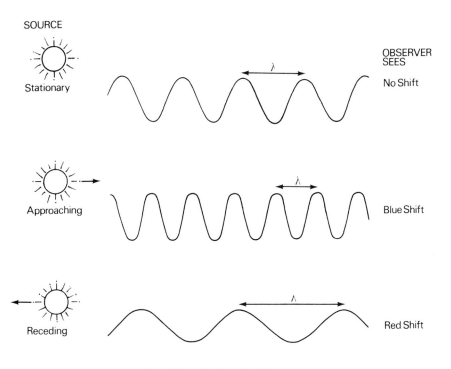

Figure 2.3 — The Doppler Effect.

FURTHER READING

Bondi, H., *Cosmology,* (Cambridge University Press, 1961).
Heath, Sir Thomas, *Aristarchus of Samos* (Oxford University Press, 1959).
Hoyle, F., *Frontiers of Astronomy* (Mercury Books, London, 1961).
Russell, Bertrand, *History of Western Philosophy* Allen and Unwin, 2nd ed.
       1961).

3

Earth Time

They have impal'd within a Zodiake
The free-borne Sun, and keepe twelve Signes awake
To watch his steps; . . . . . .

John Donne (1573–1631)
An Anatomie of the World.

3.1 SIDEREAL AND SOLAR TIME

Physical events occur in the world — a glaringly obvious remark which seems to be scarcely worth making, even though the assertion is, without doubt, true. Yet, reflect on a world in which events were absent. Nothing would change, nothing would live, nothing would die; a sameness would exist for eternity; no effect would follow cause, because there would be no cause in the first place. The word to describe such a world is timeless: it would be a world without time. Time has meaning only where there is change. Our world is not a timeless world. Everything is in a state of flux; a state of ending and becoming, stopping and starting, annihilating and creating. The sequence of events is time itself, and we need a measure of it. To describe, to plan, to predict, we require a quantification of the direct physiological appreciation of time which we all have built into us. We need a reference series of events to act as a time-rule; in other words we need a clock. The most obvious and powerful clock is the earth itself revolving about its axis.

A clock is nothing more than a sequence of events. What sequence of events is provided by the earth's rotation? There are, in fact, a large number, but two stand out with special significance. The simplest is to take the passage of a given star across the meridian, the imaginary north-south line drawn across the sky and passing directly overhead, as the definition of the tick of the earth's clock. The stars are so distant from us that, to a very good approximation, they can be regarded as being fixed in direction throughout the earth's orbit of the sun. Nor have they changed their positions much in historic time. A star rises in the east, crosses the meridian, and sets in the west. Between successive meridional transits the earth has rotated exactly 360°. The time measured in this way is known as sidereal time. A sidereal day corresponds to exactly one rotation of the earth, with respect to the 'fixed' stars.

To relate the length of the day to exactly one revolution seems the obvious and sensible thing to do. But as a measure of time for everyday life it will not do, because we are far more affecred by the sun than by the 'fixed' stars. Like the stars, the sun rises in the east, crosses the meridian, and sets in the west, but the period between successive transits is not the same as that for the stars. Besides revolving, the earth follows its nearly circular orbit around the sun. Both revolution and orbit are rotations in the same sense, and therefore the earth has to rotate a little bit more than 360° for the sun to return to the meridian, (Figure 3.1). Successive transits across the meridian of the sun beat out solar time, and the period between them is the solar day, which is about four minutes longer than the sidereal day. If we insisted on sidereal time as the basis of our civil time, we would find that a day in which the sun stood at its zenith at sidereal noon, would be followed six months later by a day in which the sun reached its zenith at sidereal midnight. This is clearly an unsatisfactory and highly disturbing way of measuring time, and so we opt for clocks which keep solar time, where a solar day corresponds to one rotation of the earth with respect to the sun.

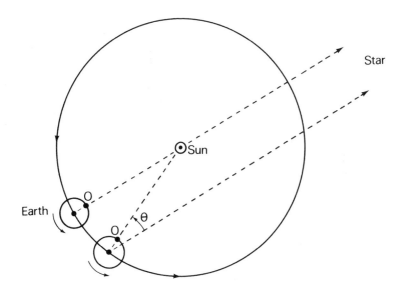

Figure 3.1 – Difference between solar and sidereal days. Θ is extra angle earth has to rotate beyond 360° to bring sun back overhead, at the point O on the earth.

Solar time is a natural measure, not only for man, but for all terrestrial life forms. Besides the daily periodicity evident in the physical environment, there are many periodicities observed in our own and other species which roughly correspond to the solar day. Such rhythms are known as circadian (literally, about a day), and they have a noticeable affect on our behaviour. Roughly, our vitality is lowest about four o'clock in the morning, and reaches its peak in the early afternoon (unless we have eaten a heavy lunch!). Anyone who has kept pets will have noticed a circadian pattern in their activity. It is scarcely a coincidence that such rhythms occur. The sequence of night and day has persisted for thousands of millions of years, and has laid its imprint on the evolution of life on the planet.

3.2 MEAN SOLAR TIME

It is one thing to appreciate the significance of solar time, and quite another to put it on a firm quantitative basis. There are many aspects of the earth's motion to be taken into account. It would be comparatively simple to define accurately a solar day of 24 hours if the earth's axis of rotation were at right-angles to the plane of the orbit, if the orbit were exactly circular, and if we did not have an outsized satellite like the moon. The absence of tilt would mean that the sun would appear to follow the same track through the sky, rising at the same point on the horizon each day, crossing the meridian at the same altitude, and setting without fail at the same point in the west. An exactly circular orbit would mean that the earth would travel at exactly the same speed around the sun, and so the time between successive noons would not depend upon where the earth was in its orbit. Elimination of the moon would get rid of its gravitational disturbances, such as ocean tides, which tend to slow the earth down.

But the earth's rotational axis is not at right-angles to the orbital plane (Figure 3.2). It is not, as it were, vertical, but tilted at the large angle of 23.45°. The axis points almost directly at the star Polaris, the North Star, and keeps pointing in the same direction throughout the orbit. For half the year it is inclined towards the sun, and the northern hemisphere experiences spring and summer, and for half the year it is inclined away from the sun, and the northern hemisphere experiences autumn and winter. The seasonal variation manifests itself by a north-south motion of the sun in the sky.

Nor is the earth's orbit circular, though it is very nearly so. At perihelion, the point of the orbit nearest the sun, the earth is 1.407×10^8 km from the sun. At aphelion, the point furthest from the sun, the earth-sun distance is 1.5207×10^8 km. There is relatively little difference but it is not negligible. Happily for the northern hemisphere, perihelion occurs during winter, and this helps to moderate the seasonal effect produced by the tilt. The reverse is true for the southern hemisphere, which in itself would suggest that the latter suffers colder winters and hotter summers than does its northern counterpart. The greater area of ocean in the south compensates this trend very effectively. This state of affairs will not always be true, as we shall see. But in addition to climatic effects, the elliptical orbit introduces a complication into our definition of solar time through the variation of the velocity of the earth in its orbit. Planets in elliptical orbits go faster when they are near the sun, and slower when they are further away. One of Kepler's laws describes this in a statement that the line joining sun to planet sweeps out equal areas in equal times. To maintain equal areas when the planet is close to the sun therefore entails that the planet move faster. The effect is small in the case of the earth. Its orbital velocities at perihelion and aphelion are 30.272 kms^{-1} and 29.278 kms^{-1} respectively. Nevertheless, it means that the earth near perihelion has to rotate further to get the sun back to the meridian, than it has to near the aphelion. The days in northern winter, as defined by successive meridional transits of the sun, are longer than they are in the summer!

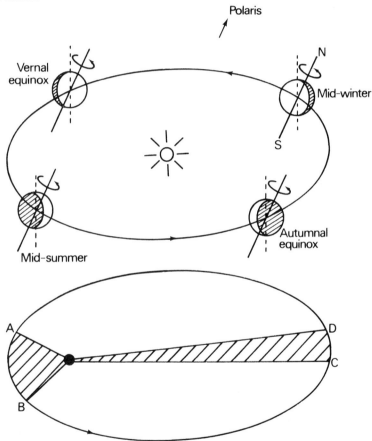

Figure 3.2 – (a) Direction of rotation axis remains substantially fixed in earth's orbit around sun. The seasons are for the northern hemisphere.
(b) Kepler's law. Exaggerated elliptical orbit with shaded equal areas. Time for satellite to go from A to B equals time for it to go from C to D.

Both the tilt of the rotation axis and the ellipticity of the orbit make the time the earth takes to go once around the sun an important period in our environment. Twice a year the sun moves across the meridian at the equator directly overhead. These are the equinoxes when day and night have equal lengths. One occurs on or near the 21st March, the other on or near the 23rd September. The period between successive vernal (spring) equinoxes is kown as the tropical year. Since it is grossly inconvenient to have a day which varies in length over the year, we perform an average, and arrive at the measure known as mean solar time. The length of the mean solar day is then divided into 24 hours. The difference between mean noon and true noon can reach about 15 mins. Having thus defined a mean solar time we can measure the tropical year, and this turns out to be 365.24219879 mean solar days, to the nearest millisecond.

3.3 THE CALENDAR

It is naturally of some importance to know how long a year is in order to construct a workable calendar. In 45 B.C. Julius Ceasar sanctioned the general use of a calendar based on having 365 days in the year, and 366 days in every fourth year. The foundation of the Julian calendar was therefore a year consisting on average of 365.25 days, which was too long by only about 0.008 days. Nevertheless, the discrepancy became gradually apparent over the centuries. By the sixteenth century it was almost two weeks: the date of the vernal equinox was slipping backwards; spring was invading the traditionally winter months. Action was finally taken in 1582 by Pope Gregory XIII. The date of the vernal equinox was defined as 21st March, and to shorten the average year the leap years in 1700, 1800, 1900, but not 2000, were to be eliminated. This makes the year of the Gregorian calendar equal on average to 365.2425 days, only 0.0003 days or 26s too long. This is our calendar today. Provided we follow this calendar and drop three leap years every four hundred years we will not run into trouble for a few thousand years.

3.4 THE SIDEREAL YEAR

The tropical year is the period associated with the earth's orbit of the sun, which has direct relevance to everyday life, since it goes in step with the seasons. It is not the only orbital period. There is the orbital period relative to the stars, namely, the sidereal year.

The plane of the earth's orbit is known as the plane of the ecliptic, because when the moon, in its orbit around the earth, crosses that plane, it may eclipse the sun or be eclipsed by the earth. If we imagine this plane stretching out from earth and intersecting the night sky of stars (the celestial sphere) it makes an imaginary line across the firmament which is very close to the track taken by the planets in their apparent journey from east to west. (Figure 3.3). This is because all of the planetary orbits tend to lie in the same plane, and so we see them in the night sky always near the ecliptic. The constellations of stars intersected by the ecliptic provide a fixed background against which planetary motions can be seen, and historically they have acquired names, (and in astrology, a considerable amount of character!). Twelve names have been given, so that each may be taken to occupy a 30° arc of the ecliptic (Figure 3.4). If we could see the constellations during the day we would see the sun entering Pisces about the vernal equinox (21st March), and in successive months enter Aries, Taurus, Gemini, Cancer, Leo, Virgo, Libra, Scorpio, Sagittarius, Capricorn, Aquarius, and then back to Pisces again. If we measure how long it takes for the sun to complete this imaginary journey we get a measure of the sidereal year. The sidereal year turns out to be 20 minutes longer than the tropical year.

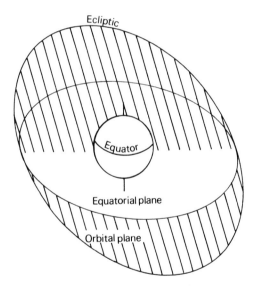

Figure 3.3a – Equatorial and orbital planes and the ecliptic, the imaginary line across the sky made by the intersection of the orbital plane and the celestial sphere, along which the sun and the planets move and where the signs of the zodiac are found.

Figure 3.3(b) (*following page*) – Tracks of solar eclipses in Europe.
(Crown copyright, Science Museum, London)

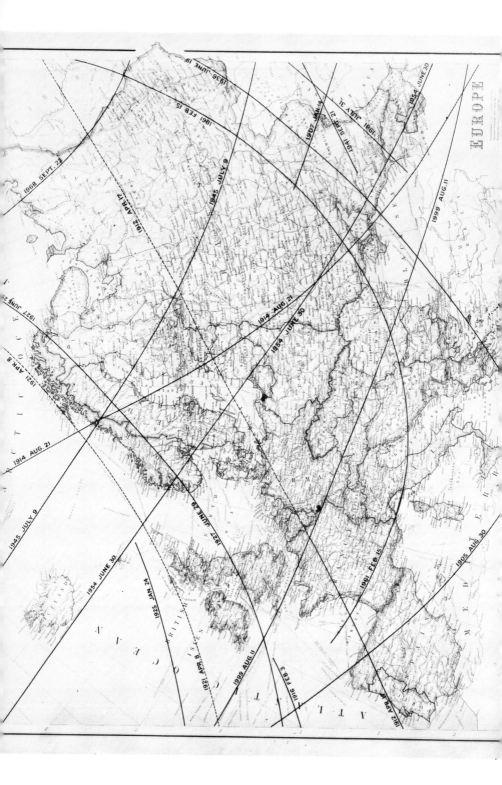

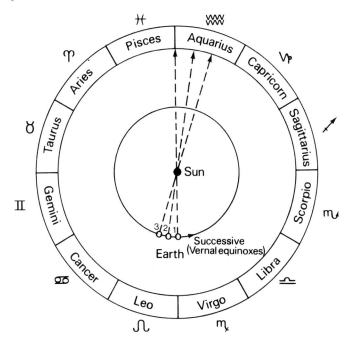

Figure 3.4 – The Zodiac and the Precession of the Equinoxes. Because the sidereal year turns out to be 20 mins longer than the tropical year, successive vernal equinoxes find the sun changing position so that the sun appears to move from Pisces to Aquarius.

3.5 PRECESSION OF THE EQUINOXES

That extra 20 min means that the point on the ecliptic reached by the sun at the vernal equinoxes is moving in a retrograde fashion from Pisces to Aquarius. The rate at which this retrograde motion is taking place, as determined by the difference between the tropical and the sidereal years is 1/25,800 of a year. This phenomenon is known as the precession of the equinoxes. Like a spinning top, the earth precesses. The direction of the rotational axis changes slowly with respect to the fixed stars. Now it points very nearly at Polaris. In 25,800 years time it will return to Polaris. In the meantime the intersection of the axis with the celestial sphere describes a substantial circle, during which there will be no well-defined north star. One consequence of precession is that the northern winter will get somewhat more severe as the winter months move to coincide with the period during which the earth is furthest from the sun.

A spinning top precesses because of the gravitational torque which tries to rotate the inclined top so that it lies flat on the ground. The torque which causes the earth to precess is also gravitational. It arises through the gravitational forces exerted by the sun and the moon, and to a much lesser extent the planets, acting on an earth, which is not quite spherical but bulges at the equator. The gravitational torque attempts to pull the plane of the equator into the plane of the orbit, but since the earth is spinning it succeeds only in causing precession.

3.6 GRAVITATIONAL TORQUES AND TENSIONS

This illuminates a general point concerning the action of gravitation. Besides the force of attraction which acts on the centre of mass there exists, in general, a torque. Furthermore there is a stretching force, a tension that tends to disrupt the body. The interesting thing about the gravitational torque and the gravitational tension is that they both obey an inverse-cube law rather than the familiar inverse-square law, and this makes nearby bodies take on much more importance than distant ones.

This can be seen most simply by looking at the forces exerted by a distant body on a rigid dumb-bell-shaped object, whose mass is concentrated equally at both ends (Figure 3.5). If the distance to the centre of mass is R, then to a good approximation one end of the dumb-bell is at a distance $(R-a)$ and the other $(R+a)$, where a is the projection of half of the length of the dumb-bell along the line to the distant gravitational source.

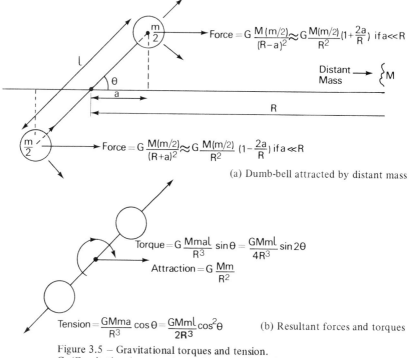

$$\text{Force} = G \frac{M(m/2)}{(R-a)^2} \approx G \frac{M(m/2)}{R^2}(1+\frac{2a}{R}) \quad \text{if } a \ll R$$

Distant Mass $\longrightarrow$ $\{M$

$$\text{Force} = G \frac{M(m/2)}{(R+a)^2} \approx G \frac{M(m/2)}{R^2}(1-\frac{2a}{R}) \quad \text{if } a \ll R$$

(a) Dumb-bell attracted by distant mass

$$\text{Torque} = G \frac{Mmal}{R^3} \sin\theta = \frac{GMml}{4R^3} \sin 2\theta$$

$$\text{Attraction} = G \frac{Mm}{R^2}$$

$$\text{Tension} = \frac{GMma}{R^3} \cos\theta = \frac{GMml}{2R^3} \cos^2\theta$$

(b) Resultant forces and torques

Figure 3.5 – Gravitational torques and tension.
G (Gravitational constant) = 6.668×10^{-11} Nm^{2}kg^{-2}.

Because the strength of gravitation falls off with distance, the end of the dumb-bell nearer the distant mass is pulled more than the other end is, and so the dumb-bell experiences a tension, a force tending to pull it to pieces. Unless the dumb-bell is aligned along the direction of the gravitational force it also experiences a torque, a twist which arises once more from the *difference* between the gravitational attractions at the ends. How does this difference depend upon how far away the distant mass is? It is easy to show that when the dumb-bell is very small it depends inversely on the cube of the distance R. Essentially, these forces are related to the *gradient* of the gravitational force, that is, to the amount the gravitational force decreases in one metre, and the gradient of an inverse-square law force always varies as the inverse-cube of the distance. The strengths of the tension and torque are, of course, bigger the further apart the ends of the dumb-bell are — the more extended the body, the more it tends to be disrupted — but for a given dumb-bell these forces are dependent on the mass M of the distant body and the cube of the reciprocal of its distance. This law holds for all bodies whose dimensions are much smaller than R, so although the earth does not look particularly like a dumb-bell (on the other hand, because of its being shaved off at the poles it could be thought of as a *negative* dumb-bell) it reacts to those forces as if it were.

Gravitational torques and tensions exerted on a particular object are therefore proportional to (M/R^3). This quantity is calculated in Table 3.1 for the moon and planets at their closest approach to the earth, and expressed as a proportion of that of the sun at perihelion. Because of the inverse-cube law the influence of the moon is even bigger than that of the sun, in spite of the fact that the moon's mass is relatively tiny. The influence of the planets is practically insignificant.

Table 3.1 — Strengths and Gravitational Torques and Tensions at the Earth, of the Moon and Planets Relative to the Sun

Body	Mass (in earth masses)*	Nearest Distance (km) (neglecting orbital obliquities)	Strength Relative to the Sun
Sun	3.33×10^5	1.47×10^8	1.00
Moon	1.23×10^{-2}	3.63×10^5	2.45
Mercury	5.32×10^{-2}	7.72×10^7	1.10×10^{-6}
Venus	8.17×10^{-1}	3.82×10^7	1.40×10^{-4}
Mars	1.07×10^{-1}	5.45×10^7	6.29×10^{-6}
Asteroids	2.84×10^{-1}	4.14×10^8 (mean)	4.36×10^{-8}
Jupiter	3.18×10^2	5.88×10^8	1.50×10^{-5}
Saturn	9.51×10^1	1.20×10^9	5.24×10^{-7}
Uranus	1.45×10^1	2.59×10^9	8.41×10^{-9}
Neptune	1.72×10^1	4.31×10^9	2.05×10^{-9}
Pluto	1.80×10^{-1}	4.29×10^9	2.20×10^{-11}

*The mass of the earth is 5.977×10^{24} Kg.

Gravitational influences on earth of extra terrestrial bodies are therefore associated with the gravitational gradients of the sun and moon. Before going on to discuss how these influences affect the length of the solar day, let us briefly make a point concerning gravitational tension. The latter will rise rapidly if a satellite approaches its planet too closely. Nearer than a certain radius, known as the Roche limit, the tension overcomes the gravitational adhesion of the satellite and the body disintegrates. For the moon the Roche limit is about 2.89 Earth radii, well within the present radius of the moon's orbit. The general point worth empahsizing is that gravitational forces exerted on a body are disruptive. However, in the case of small bodies, such as artificial satellites or asteroids consisting of single lumps of rock, such forces are of no importance because of the large mechanical strength of homogeneous solids.

3.7 TIDAL FRICTION

On the earth the most obvious effect of these forces is the tidal motion of seas and oceans. Tension tends to pile water in heaps on opposite sides of the earth's surface, and any point on the shore experiences two high tides and two low tides a day. Twice a month the sun and the moon are aligned and extra-high tides, the spring tides occur. Weak tides, the neap tides, happen twice a month when the pulls of the sun and moon are more or less at right-angles to one another. Actually the daily period is about 50 minutes longer than 24 hours because the moon, orbiting the earth in the same direction as the earth's rotation, crosses the meridian that amount of time later each day. The sidereal period for the moon to go once round the earth is 27.3217 days, but the synodic period, the time between successive new moons, is longer for the same reason that the' solar day is longer then the sidereal day: the earth and moon orbit the sun (Figure 3.6). The tides follow the synodic period, which is 29.5306 days.

The size and timing of tides are naturally dependent upon local conditions, and no seaman would rely on the general remarks made above but would consult an accurate tide-table. In open ocean the tidal range is less than one metre, but in shallow seas and estuaries it can exceed 10m, and the influence of coast-lines on the phase of the tide relative to the positions of the sun and moon is very large. Tides are also affected by the ellipticity of the moon's orbit, by the inclination of the orbit to the plane of the ecliptic, and by its precession. At perigee, the closest approach to the earth, the moon is at a distance 3.63×10^5 km; at apogee, 4.06×10^5 km. Its orbit lies in a plane inclined $5.13°$ to the plane of the elliptic, and the whole orbit precesses in space in a retrograde manner with a period of 18.5 years. Accompanying this is a direct, rather than retrograde, precession of the major axis of the elliptical orbit in the orbital plane with a period of 9 years, and indeed several other irregularities of motion. In the earth-moon system we have a real, physical dumb-bell, albeit with very dissimilar masses, which rotates about a common centre-of-gravity and experiences pre-

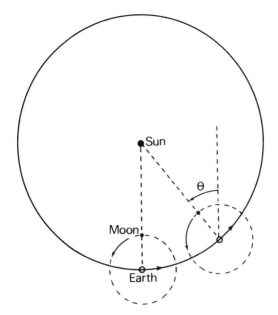

Figure 3.6 — Synodic period, the time between new moons. Θ is extra angle through which moon has to orbit from 360° rotation around the earth.

cessional torques exerted by the rest of the solar system. The movement of our satellite is therefore very complex and is a study in its own right. But there is yet another factor which determines the magnitude of the tidal effect, and this is the phenomenon of earth tides, tides in the solid crust. Slight deformations of shape occur in response to the gravitational tension. These earth bulges reduce the relative height of the open oceanic tides by as much as a third, but because they are generally out-of-phase with shallow-sea and estuary tides, earth tides have little effect on tidal range around coast-lines.

The effect of tides on the earth as a whole is to slow down the speed of rotation. Because the earth rotates it tends to pull the tidal bulges around with it (Figure 3.7). Like the dumb-bell, the earth then experiences a gravitational

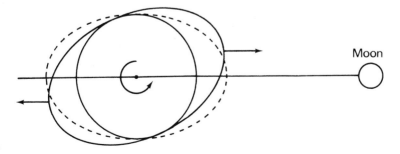

a

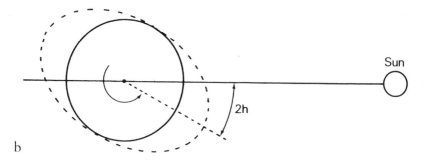

b

Figure 3.7 – (a) Tides and retarding gravitational torque. Dotted line represents tides on non-rotating earth. Earth's rotation tries to carry tides around. (b) Atmospheric tides and accelerating gravitational torque.

torque, as well as the tension which produced the tides in the first place. This torque acts, on the whole, to retard rotation. Tidal friction therefore tends to lengthen the day.

3.7.1 A longer day?

But is the day actually getting longer? This is not an easy question to answer. From existing records of eclipses and of the movement of the planets left by the Babylonians and Greeks, it was deduced a couple of centuries ago that the day appeared to be lengthening at a rate of 2ms per century and that the moon appeared to be moving away. The German philosopher, Immanuel Kant (1724-1804), was the first to suggest that tidal friction would indeed cause a lengthening of the day. But even with accurate atomic clocks the retardation rate is far too small to measure directly against the background of much larger annual fluctuations.

Theoretical estimates based on observations by satellite of the phase lag of the tidal bulge yield a figure of about 1.6ms per century. Evidence in support of this retardation, and that it has been in progress for millions of years, is shown by a study of fossil corals which lived in Devonian times, 370 million years ago. Daily growth rings, about 10 μm thick, can be seen in skeletons within broader annual bands. The number of such rings in an annual band is 400. Since tidal friction does not affect the length of the year, this observation suggests that the Devonian year consited of 400 days, and that therefore the day was 22 hours long. This figure agrees with a retardation rate of 2ms per century.

A moving away of the moon would be intimately associated with the retardation of the earth's rotation through the conservation of angular momentum. To a first approximation, which neglects the tidal influence of the sun, the moon and earth form an isolated system in which the total angular momentum must remain constant. Therefore if the angular momentum of the rotating earth is decreasing, it must have been transferred to the moon. Tides raised by the earth on the moon have slowed the moon's rotation about its own axis to the point where it presents the same face to the earth at all times. The transfer of angular momentum cannot therefore act to rotate the moon faster. The transfer has to go to the moon's orbital motion. The moon orbits faster and therefore moves away from the earth. The rate of separation would be such that if extrapolated backwards in time for 3.5×10^9 years, the moon would be only of the order of 6,000km away, well inside the Roche limit! (This raises interesting questions regarding the origin of the moon, incidentally.) An extrapolation forwards in time (some 10^{10} years hence) suggests that the rotational period of the earth will lengthen to a staggering 47.4 days, by which time the earth will present the same face to the moon at all times. That is not the end of the story, because tidal friction by the sun will continue to slow the earth. Relative to the moon the earth will now appear to rotate in the opposite direction, and consequently the moon will begin to accelerate the earth; the sign of the effect will have changed and the moon will begin to spiral slowly inwards towards the Roche limit.

In any case, the lengthening of the day may simply not occur. Tidal retardation is opposed by another phenomenon we have not yet mentioned, which acts to speed up the earth's rotation. This phenomenon is a tidal effect in the atmosphere, manifesting itself as a pressure maxima 2 hours before mid-day and 2 hours before midnight. Lord Kelvin (1824–1907) suggested correctly that these atmospheric tides were thermally induced and were thus in phase with the sun. As a consequence the sun exerts a gravitational torque on these atmospheric 'bulges' which acts to accelerate the earth's rotation. Conservation of angular momentum in the sun-earth system then predicts a movement of the earth towards the sun.

The existence of two opposing tidal torques, one associated with the sea, the other with the air, is not in doubt. The question is what are their magnitudes? It is tempting to conclude that on the evidence of the Devonian corals oceanic tides exert the stronger torque and the net effect is a lengthening of the day. Though that conclusion may have been valid 370 million years ago, it may not be valid to extrapolate that conclusion to the present day. The reason for this lies in the possibility that atmospheric tides have grown bigger as the day has lengthened owing to a resonance of solar heating with the natural frequency of vibration of the atmosphere. It is possible that a rotation period of 24 hours is exactly the right amount for both torques to be the same strength. On this view the day will never be longer than it is now.

3.8 STANDARDS OF TIME

A secular change in the length of the day draws attention to our definition of time based upon the earth's rotation. The mean solar day is not a constant, unless we define it as that in one particular year. This was, in fact, done to define the so-called Ephemeris Time. In this standard of time, 24 hours became the length of the mean solar day in the year 1900. This formed the basis of time measurements until 1972.

As the basis for a scientific standard of time it was rather unsatisfactory. The fact of the matter is that the earth does not make a good clock. Its motion is far too complex. Besides the complications associated with the tilt, the ellipticity of the solar orbit, and the axial precession, there are several other minor peculiarities. The plane of the orbit slowly oscillates causing the angle of tilt to vary between 24.60° and 21.98°, so that the precessional cycle is not quite closed. The cycle is 40,000 years. At present the angle of tilt is decreasing. Nor is the axial precession smooth, but suffers from a wavy motion called nutation, caused by the constantly turning plane of the moon's orbit, the period of which is 18.5 years. In a long cycle of 92,000 years the shape of the earth's orbit around the sun changes from nearly circular to more elliptical and back again. There is also a 6 month oscillation caused by the slight asymmetry of the northern and southern hemispheres. The obliquity of the moon's orbit produces a similar but much smaller effect, this time with a period of about 14 days. As if this were not enough, the earth wobbles itself without any help from outside. It is called the Chandler wobble, after its discoverer, and arises because the rotation axis does not lie exactly along the axis a greatest moment of inertia. Relative to the fixed stars the rotation axis remains substantially constant during the wobble, but relative to the earth's surface the pole of rotation describes a circle around the moment-of-inertia axis, with a period of between 430 and 435 days. As a result there is a cyclic variation of latitude of up to 0.14 second of arc.

Some of the important cycles in the environment are summarized in Table 3.2.

It is scarcely surprising that our standard of time is no longer based upon the earth's rotation. Since 1st January 1972, the standard second has become 9,192,631,770 periods of the microwave radiation emitted by caesium (Cs^{133}) vapour.

Table 3.2 — Cycles in the Environment

1. THE EARTH

Tropical year	365.242199 days
Sidereal year	365.25636556 days
Perihelion to perihelion	365.25964134 days
Chandler wobble	about 430 days
Magnetic pole rotation	about 7,200 years
Perihelion cycle	20,940 years
Precession of the equinoxes	25,784 years
Ice age period	about 40,000 years
Obliquity-of-ecliptic period	40,000 years
Orbital-shape period	92,000 years
Magnetic field reversals	about 200,000 years
Ice age eras	150–250 million years
Continental drift	about 400 million years

2. THE MOON

Sidereal month	27.32166140 days
Perigee to perigee	27.5545505 days
Synodic month (new moon to new moon)	29.5305882 days
Eclipse year (retrograde motion of node)	346.620031 days
Moon's orbital year (direct rotation of orbit)	411.77 days
The Saros (Babylonian cycle of eclipses = 223 lunations)	18.0300 years

3. THE SUN

Sidereal period of rotation at latitude 17°	25.38 days
Synodic period of rotation at latitude 17°	27.275 days
Sunspot period (average)	11.04 years
Sunspot-maximum cycle	about 80 years
Thermal time-constant	about 10 million years
Cosmic year (Sun's rotation around galaxy)	245 million years
Nuclear fuel time-constant	about 10 thousand million years

		4. THE PLANETS		Rotation Period (days)
Planet	Symbol	Sidereal Period (days)	Synodic Period (days)	
Mercury	☿	87.969	115.88	87.97
Venus	♀	224.701	583.92	243.0
Earth	⊕	365.256	—	1.00
Mars	♂	686.980	779.94	1.03
Jupiter	♃	4332.589	398.88	0.41
Saturn	♄	10759.22	378.09	0.43
Uranus	♅	30685.4	369.66	0.45
Neptune	♆	60189	367.49	0.63
Pluto	♇	90465	366.73	0.27

Great Years (planets return to original positions with respect to the stars).

1) Philolaus (5th century B.C.) 9^3 lunar months, about 59 years.
   (Only works for Jupiter and Saturn, Jupiter performing 5 and Saturn 2 rotations).
2) Plato (c400 B.C.) $3,600^2 = 12,960,000$ days $= 35,483.30$ years.
   (Actually Plato's Great Year is 36,000 years, each year containing 360 days).
3) Modern (author's calculations) 35,348.13 sidereal years.
   (Number of sidereal rotations for Mercury 146768.92, Venus 57459.09, Mars 18794.02, Jupiter 2,980.00, Saturn 1200.00).
4) Alignment of Venus, Earth, Jupiter and Saturn.
   (conjunction or opposition): Solar Tide Cycle 11.09 years.

FURTHER READING

Allen, C. W., *Astrophysical Quantities,* 3rd ed. (University of London, Athlone Press, 1973).

Fundamental Physical Constants, (National Physical Laboratory 1974).

Jeffreys, H., *The Earth,* 2nd ed., (Cambridge University Press, 1928).

Rudaux, L., and De Vancouleurs, G., *Larrouse Encyclopaedia of Astronomy,* (Batchworth Press, London, 1959).

Stacey, F. D., *Physics of the Earth,* (Wiley, 1969).

Toulmin, S. and Goodfield, J., *The Discovery of Time* (Harmondsworth, 1967).

Whitrow, G. J., *What is Time?,* (Thames and Hudson, 1972).

4

The Structure of the Earth

From Ymir's flesh was the earth shaped,
From his blood the salt sea,
The fells from his bones, the forests from his hair,
The arching sky from his skull;
From his eyelashes the High Ones made
Middle-Earth for men,
And out of his brains the ugly tempered
Clouds were all carved.

The Lay of Grimnir, from the Elder Edda,
translated from the Icelandic by Paul B. Taylor and W. H. Auden.

4.1 SHAPE AND GRAVITY

The earth is almost, but not quite, spherical. The cross-section revealed by a cut from pole to pole would be an ellipse rather than a circle, but a similar cut through the equator would be circular. Technically, the earth is an oblate spheroid, the shape of a squeezed tennis ball. The oblateness is very slight. In the idealized model in which the earth's shape is exactly spheroidal, the distance from points on the equator to the centre is 6,378.17 km, which is to be compared with the corresponding distance to one of the poles, which is 6,356.79 km. The difference is less than 22km. On the real earth surface, irregularities, such as mountains and trenches, account for variations of several kilometres. The plateau in Antarctica at the South Pole, for example, is about 2.8km above sea-level, and mountains near the equator, such as Mount Kilimanjaro, are over 5km high.

Naturally, the oblateness appears in measurements of circumference. In our ideal model the length of an equatorial quadrant (one quarter of the circumference) is 10,018.81km, while the length of the meridional quadrant (from pole to equator) is 10,002.02km. The closeness of the latter figure to a round and simple 10,000km is no coincidence, for the metre was first defined as the ten millionth part of the terrestrial quadrant, running from the North Pole to the equator via Paris. Now it is defined in terms of the wavelength of light emitted by the rare gas krypton (Kr) and will soon be redefined in terms of the caesium second and the velocity of light, once the latter can be determined with sufficient accuracy. Nevertheless, its historical origins make the round 40,000km for the circumference of the earth an easily remembered, and very good approximation.

The cause of the oblateness is the rotation of the earth — bodies tend to fly off the equator, but do not do so at the poles. The universal property of inertia, whereby matter attempts always to move with uniform velocity in a straight line, exerts its influence on the rotating earth, which is to some degree plastic, and deforms it from a perfect spherical form. We describe the effect from the standpoint of an observer rotating with the earth by inventing a convenient fictional force, the centrifugal force, which acts to accelerate matter radially away from the axis of rotation, to describe this inertial tendency. The effect becomes very evident when we measure the acceleration of a falling body as a function of latitude. We find that bodies fall more rapidly at the poles than at the equator.

4.1.1 Gravity and Anti-gravity

Think of the forces acting on each one of us. There is the ever-present downward pull of the earth's gravitational field, and there is our tendency to proceed happily at uniform speed in a straight line. Since the earth rotates and since we partake of this motion, the tendency to travel at uniform speed in a straight line − in a word, our inertia − tries to project us into space. We try to travel outwards along a tangent, but gravity stops us doing that. From the standpoint of the rotating earth it appears that some force, directed away from the rotation axis, is trying to project us upwards along a path inclined towards the equator. This is the centrifugal force, something familiarly experienced by anyone who has driven quickly around a bend in the road.

The earth's centrifugal force can be conceived of as having two components, one directed vertically, the other directed horizontally towards the equator (Figure 4.1). The vertical component acts against the downward force of gravity and makes the effective gravitational pull weaker than it would be in the absence of rotation. The magnitude of this vertical component is zero at the poles and maximum at the equator, and so effectively gravity is strongest at the poles and weakest at the equator. The difference is very small − the ratio of pole gravity to equator gravity is 1.0034 − but the acceleration of the effective gravity, g, can be measured to within one part in 10^8, so the rotational effect is easily detectable. Accurate measurement of g can reveal gravitational anomolies associated with local variations in the density of the earth's crust, so it is vitally important to know how g ought to vary with latitude so that gravitational surveys can be interpreted properly. The slightly increased gravitational force at the poles has produced the oblate spheroidal shape of the earth, and this departure from perfect sphericity has produced second-order changes in the pattern of effective gravitational field, all of which have to be taken into account to get an accurate theoretical picture.

The net vertical force of gravitation and rotation presses us downwards and is opposed (unless we are falling) by the pressure of material beneath us. But what about the horizontal component of the centrifugal force which tries to slide us into the equator. This is opposed by a tiny amount of gravitation which appears as a horizontal component consequential on the oblateness of the earth.

So we may stand still on the earth's surface and feel no net force. Once we begin to move, however, the delicate balance of gravitational and inertial forces is immediately destroyed, and a new effect appears – a tendency to move at right-angles to the direction of motion. This is the Coriolis effect, and we will have more to say about this in Section 7.6 when we discuss the motion of things on the earth.

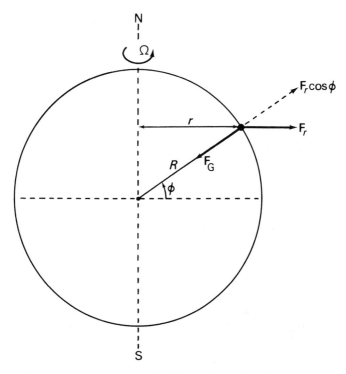

Figure 4.1 – Gravitational and centrifugal forces. Gravitational force on unit mass on surface $= G\dfrac{M}{R^2}$. Centrifugal force on unit mass $= \Omega^2 r$ $= \Omega^2 R\cos\phi$ where $\Omega =$ angular frequency (radians per second) of earth's rotation, and $\phi =$ latitude.

As regards the vertical forces which act upon us, it is apparent that the centrifugal component is a sort of 'anti-gravity'. If we reduce it by travelling westwards we become heavier, and we get lighter if we travel towards the east. The faster we go east the lighter we get. If freight charges do not appear to take this fact into account, the early launches of satellites did. Anti-gravity is strongest at the equator so the easiest way to get a satellite up is to launch it eastwards into an equatorial orbit. There exists a velocity which makes anti-gravity as strong as gravity on the earth's surface. That condition of weightlessness occurs for a velocity of 7.9kms^{-1} (about 28,000 km hr^{-1}), which is considerably greater than the equatorial rotational velocity of about 0.5kms^{-1}, so one gains a 6% advantage by setting off eastwards rather than northwards or southwards. It is relatively a small step to reach a velocity which will take us out of the earth's gravitational field altogether. This velocity, known as the escape velocity, is 11.2 kms^{-1} (about 40,000km hr^{-1}, i.e. once around the earth in 1 hr). We will return to the topic of escape velocity in Section 7.5 when we consider the environmental influence of gravitation further.

4.2 SEISMIC WAVES

Exploration of the earth's crust by gravity surveys, or magnetic field surveys, shows up local features only. To explore the deep internal structure of the earth we need a farther-reaching probe than that provided by field surveys. Such a probe is provided naturally by earthquakes, which radiate seismic waves throughout the globe. Seismic waves travel outwards from the epicentre of an earthquake and may be detected over huge areas of the earth's surface by the minute vibrations they cause in sensitive detectors called seismographs. Many such waves travel right through the earth, and the study of the time they take to do so, and the pattern of intensity they make over the earth's surface, enable us to form a good idea of the internal structure of our planet.

Seismic waves are no different from the elastic waves that are found in all solids. Two sorts of waves can travel through the bulk of a solid (Figure 4.2). If the atoms and molecules which make up the material vibrate in a plane at right-angles to the direction of propagation − as the electric and magnetic field does in an electromagnetic wave − we call the wave transversely polarised. Since some parts of the solid move up as other parts move down, transversely polarised waves are often called shear waves. The other sort of wave is essentially an acoustic wave, consisting of alternate regions of compression and rarefaction. The motion in this case is directed backwards and forwards along the direction of propagation, and the wave is said to be longitudinally polarised. Besides the bulk waves, there are two sorts of waves which travel along the surface of a solid, which are again distinguished by the direction in which particles oscillate. If the oscillation is transverse and entirely in the surface plane we call the disturbance a Love wave, named after its discoverer one hastens to add. Love waves can propagate only if the vibrational properties vary with depth. The other type of wave, again named after its discoverer, is known as a Rayleigh wave, in which the particle motion contains a component perpendicular to the surface. (In uniform solids the Rayleigh wave is the only type of surface wave which can propagate.) One can think of surface waves as waves which have become trapped between the air above and the dissimilar material below by being continually reflected from the boundaries, in much the same way that a beam of light propogates along a glass 'optical fibre'.

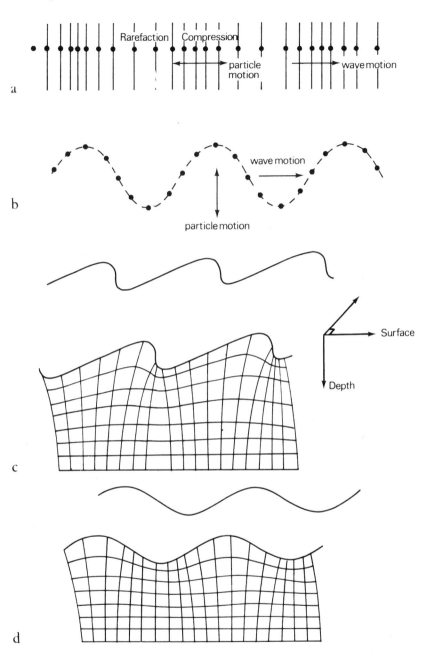

Figure 4.2 – (a) Longitidinally-polarised wave. (b) Transversely-polarised wave. (c) Love wave. (d) Rayleigh wave.

4.2.1 Elastic waves

The velocity of an elastic wave is governed by two factors. One factor is the size of the elastic restoring force which opposes the deformation of the material, and the other is the inertia of the particles forced into oscillation. In elastic processes it is never the size of the deformation in itself which is important: it is the strain. In the case of a linear deformation, strain is defined as the change in length per unit length, e.g. if two atoms in a solid were 2 Å (1 Ångstrom $= 10^{-10}$m) apart in unstrained material and 2.2 Å apart in strained material, the strain would be 0.1 (which is, incidentally, an enormous strain — most strains in elastic processes in solids are much less than 10^{-4}). Shear strains are defined as changes in angle (twist) per unit angle and volume strains are defined as changes in volume per unit volume. In elastic waves, patterns of stress and strain propagate through the material. For a given wavelength, the greater the restoring stress generated by a unit strain, the faster the particles will oscillate, and hence the higher the frequency. This effect will be enhanced if the particles have a small inertia. Since the velocity v of a wave is related to its frequency f, and wavelength λ by the equation $v = f\lambda$ a high frequency for a given wavelength entails a high velocity. The measure of restoring stress (force per unit area) generated by unit strain is the elastic constant c, sometimes known as the elastic modulus, and the measure of inertia is the mass density ρ. The theory of elastic waves gives for the velocity the expression $v = \sqrt{(c/\rho)}$. We can see that this expression is dimensionally correct. The elastic constant is a stress — a force per unit area. The latter is essentially an energy density i.e. energy per unit volume, because force times distance is energy and area times distance is volume. But so is the quantity ρv^2: it is twice the kinetic energy of a unit volume of the material travelling at the wave-velocity. But the material itself does not travel at the wave velocity: only the pattern of stress and strain does that. The material merely oscillates. So, taking care not to put too much physical significance on these energy densities, we can see that the equation for the wave velocity is dimensionally correct and plausible.

Solids can resist pushes and pulls more strongly than they can resist twists. The elastic constant for the former type of deformation is Young's modulus, and for the latter type is the shear modulus. Young's modulus is always the greater, and so for a given material longitudinally polarised waves travel faster than transversely-polarised waves. One vitally important difference between the two types of bulk wave is that shear waves cannot propagate through a liquid to any great extent, whereas compressional waves can. Liquids·cannot counter shears, but they can resist being squeezed. This fact has enabled us to deduce the existence of the earth's liquid core. Furthermore, since the velocity is a function of elastic constant and density, a study of the propagation speed of seismic waves through the earth enables us to deduce the variation of density, temperature and composition with depth. Where these quantities suddenly change there will be reflection and, if the wave can propagate, refraction. If the changes are gradual a continuous refraction occurs and the waves in general follow curved paths.

4.2.2 Vibrations of the earth

Seismic waves are therefore indispensable probes into the deep structure of the earth. The two bulk waves are known as primary waves, or P-waves, and secondary waves or S-waves. P-waves are longitudinally polarised and travel (near the surface) with an average velocity of 8.11 kms^{-1}, and S-waves are transversely polarised, propagating (near the surface), with an average velocity of 4.34 kms^{-1}. The velocities of Rayleigh and Love waves are respectively 5.60 kms^{-1} and 3.40 kms^{-1}. P-waves, travelling right through the twelve thousand kilometres of the earth, take about 25 minutes to reach the other side. They have to be detected against a continuous background of other vibrations, for the earth is never still. There are waves, for instance, of period about 12 hours associated with tides which produce earth strains of the order of 10^{-8}. And there are a multitude of smaller vibrations, known as microseisms, corresponding to strains of the order of 10^{-10}, produced by a host of causes ranging from waves breaking on the shore, to the traffic on a motorway, and trees swaying in the wind. Another seismic effect is the excitation of free oscillations of the whole earth by major earthquakes, which makes the earth ring like a bell (but at too low a frequency to hear). All these vibrations are worthy of investigation in their own right, but here we will limit ourselves to the results obtained from arrays of seismometers all over the earth's surface, monitoring the arrival of seismic waves from many individual earthquakes over many years.

4.3 INTERNAL STRUCTURE

The pattern of arrival of P and S waves over the surface of the earth following a large earthquake is extremely revealing (Figure 4.3). Suppose the earthquake is at the north pole. Both types of wave are detected at latitudes north of latitude 15°S, but between this latitude and 52°S, there is a shadow zone in which only a small amount of P-wave activity is observed, and scarcely any S-wave activity. From 52°S to the south pole P-waves once more appear but no S-waves.

This pattern is indicative of a sizeable liquid core which prevents the passage of S-waves, and reflects P-waves, so that a shadow zone exists. The small amount of P-wave activity in the shadow zone suggests that this is caused by the reflection of P-waves off an inner solid core. We thus have a three-fold division of the earth: (1) an inner solid core, (2) a liquid core, (3) an outer, solid mantle. (Figure 4.3 and Table 4.1).

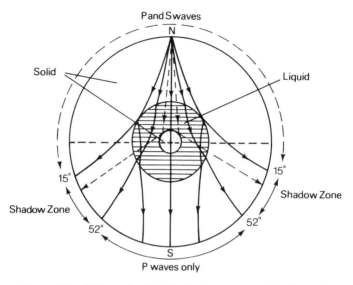

Figure 4.3 – Pattern of arrival of seismic waves over globe for earthquake at North Pole. Trajectories are those of P-waves.

The mantle does not, however, extend to the surface. Observations of P- and S-waves produced by explosive charges detonated beneath the surface, show that strong reflections occur which indicate a discontinuity some 30km below the surface. This is known as the Mohorovicic discontinuity (the Moho for short), named after its Yugoslav discoverer. It marks the division between the underlying mantle and the uppermost layer of the earth called the crust. The crustal layer, therefore, forms a fourth division of the earth (Figure 4.4).

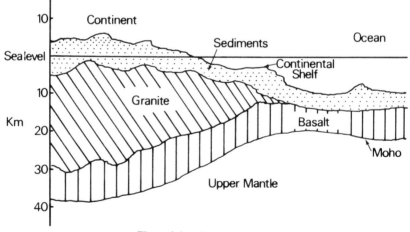

Figure 4.4 – Crustal structure.

Interestingly, the Moho is found to be less deep under the oceans than it is under the continents. Immediately above the Moho there is a 5 km thick layer of basalt, a fine-grained rock which occasionally surfaces in lava flows associated with volcanic activity. Underneath continents, but missing underneath oceans, is a 20 km thick layer of granite, a coarse-grained rock consisting of quartz, feldspar and mica crystals, which is characteristic of many mountain ranges. Generally overlying the granite in continents and the basalt in the oceans are various kinds of sedimentary rocks, such as limestone and sandstone, conglomerates and plain mud and clay, which together form the surface sediments. Naturally large variations of thickness occur, but the distinction between continent and ocean is striking. The continents are huge blocks, essentially of granite, with roots which stretch down to the Moho some 40 km beneath the surface. The oceanic crust is relatively thin, essentially basalt, with the Moho only some 5km deep. Because the mantle is dense and approaches closest to the surface under the oceans, gravity is significantly stronger over the oceans than it is over the continent.

Table 4.1 – The Internal Structure of the Earth

	Feature	Rough Maximum depth (km) Continents	Oceans	Density (gcm^{-3})	Temperature (K)
Crust {	Surface sediments	2	2		
	Granite layer	25	–	2.64	$2 \times 10^{-4} Kcm^{-1}$
	Basalt layer	33	10		
	Moho	33	10		
	Mantle	2,900		3.4 to 5.5	700 to 4,300
	Liquid core (liquid iron)	5,000		9.5 to 11.7	4,300 to 6,000
	Inner core (solid iron)	6,370		16.0 to 17.3	6,000 to 6,400

Table 4.1 summarises the variation of density and temperature with depth, as obtained from the analysis of seismic-wave propagation. Both quantities increase. That the density increases with depth is understandable, both from the fact that lighter material will tend to 'float' upwards, and from the fact that gravitational compression becomes more intense. The temperature rise has a less obvious origin. If the earth's only source of heat were the sun one would expect temperature gradients to be determined by latitude rather than mere depth. The pattern suggests that the earth has its own reservoir of heat. At one time it was assumed that the earth was formed in a hot condition and is still cooling, but for

reasons which we will go into in a later chapter, it is now thought that a fiery beginning is unlikely. Nevertheless, some of the heat generated during the earth's formation may still remain. Whatever its origins were, the earth does, in fact, possess an active heat source. It is provided by radioactivity. The radioactive decay of atoms, principally isotopes of uranium (U), thorium (Th), and potassium (K), which are especially concentrated in the crustal rocks, provide a slow but steady production of heat, insignificant as far as determining the surface temperature goes, but capable over thousands of millions of years of heating up the interior of the earth. We live on the surface of a dilute, but well-insulated atomic pile, and reminders of that fact are the various hot-spots, such as geysers and volcanoes. It is interesting to know that the centre of the earth is as hot as the surface of the sun.

4.4 EARTH'S MAGNETIC FIELD

The physical conditions deep inside the earth are far outside anything we can reproduce in the laboratory. There are obvious dangers in extrapolating the properties of materials in normal conditions to provide expectations of behaviour when the material is subject to the high pressures and temperatures of the core, but this is what we have to do. Thus we infer that whatever composes the liquid core becomes so crushed by the enormous pressures near the centre that it solidifies even though the temperature is so high. Because many meteorites are abundant in metallic iron, it is widely accepted that this element is the main constituent of the core. This view is also consistent with theories of astrophysics concerning nuclear processes and evolution of stars. The special properties of the nucleus of iron means that this element is destined to be a major component of the ash of burnt-out giant stars. It is likely, therefore, to have formed an appreciable part of the debris of some supernova explosion, out of which our solar system formed. A special attractiveness of the view that the core consists of iron is that it can help to explain the existence of the earth's magnetic field.

Where the magnetic field comes from remains something of a mystery. Its origin and somewhat erratic behaviour are topics of considerable interest at the present time. The field itself, as measured at the earth's surface, is broadly what one would expect if there were a vast electric current circulating more or less in the equatorial plane of the core around the rotation axis (Figure 4.5). Such a field is called a dipole field, since it is like the field produced by a bar magnet. Iron is a good conductor and even the rocks deep inside the mantle, where the temperature is high, will act like a semiconductor, so the existence of electric currents is certainly a possibility. There are also batteries, such as the junction between the liquid core and the mantle, and the radial temperature gradient, which could initiate a separation of charge or generate thermoelectric currents. But these effects on their own cannot plausibly explain the curious behaviour of the field.

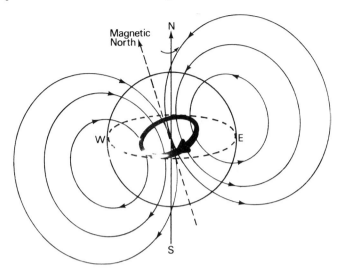

Figure 4.5 – Earth's magnetic field. Current required in single loop is 4.4×10^9 amperes flowing in opposite sense to rotation.

To begin with, the axis of the field does not lie along the rotation axis, but at an angle, known as the declination, to it. At present the declination is about $11°$, and the strength of the horizontal component at the geomagnetic equator is about 3.1×10^4 gammas (0.31 gauss). The centre of the field is not, as one might expect, at the centre of the earth, but displaced by some 300 km. Nor do any of these quantities remain fixed. At the beginning of the 19th century the declination was over $20°$. The magnetic north pole wanders around the geographic axis with a westerly precession of $0.05°$ of longitude a year. At this rate it will go once round in 7,200 years. The strength of the field is decreasing by 0.05% every year and so, at this rate, it will be very small indeed after its round trip. To complicate matters more, various magnetic anomalies are observed over the earth's surface, in which the field differs from the dipole field by as much as 30% and some of these almost certainly, have to be caused by deep-seated processes in the liquid core.

Besides having other, less striking, eccentricities which we have not mentioned, the earth's magnetic field has the remarkable property of reversing itself. Evidence for this comes from paleomagnetism, which is the study of fossil magnetism in rocks. Most rocks contain a small percentage of magnetic minerals, which are minerals based on magnetite (Fe_3O_4) or hematite (Fe_2O_3). These oxides of iron become permanently magnetised when they cool in lava flows or are deposited as sediments in the presence of a magnetic field, with the result that a record of the direction of the field is frozen in. A study of fossil magnetism in many volcanic formations and in sedimentary rocks in both hemispheres show distinct evidence of field reversals, and radiometric dating of these formations give the time scale. Definite epochs occur during which the polarity of the field remains more or less constant, with an epoch lasting for about 200,000 years, although a few shorter time reversals are also observed. An estimate of the time during which the field flips over is difficult to get at with any accuracy, but it is assessed at about 5,000 years. One vitally important consequence of reversals is that during the transition from one polarity to the other, when the field becomes weak and perhaps disappears, the earth loses its magnetic screen which shields it from cosmic radiation. This leads to a greatly enhanced mutation rate which may have caused certain species to become extinct and others to evolve rapidly.

In view of the idiosyncratic behaviour of the earth's magnetic field it is not surprising that we have no completely successful theory. The processes which give rise to the field are clearly not going to be simple. It is, however, widely accepted that the answer lies in the magnetohydrodynamics of the rotating fluid core. Several theoretical models can be constructed to show that a spherical body of conducting fluid with suitable internal motions can act as a self-exciting dynamo, producing persistent currents and an associated magnetic field. Basically what happens is that any stray magnetic field will induce a current in a rotating conductor, and this current will produce its own magnetic field, which can induce further currents in neighbouring rotating conductors. If the magnetic field produced by these currents reinforces the intial magnetic field we have the necessary positive feedback to make the processes self-generating. It is interesting that even the simplest models show that the self-excited field oscillates about a mean value and occasionally reverses spontaneously, which is rather like the behaviour of the earth's field.

4.5 MANTLE CONVECTION

A vertical temperature gradient in a fluid produces convection. Hot fluid expands and rises and cold fluid becomes dense and sinks. Convection cells develop in which the fluid slowly rotates in a vertical plane, transporting heat by being itself transported. It seems at first sight that such an effect could not possibly take place in a solid, but in fact it can. In response to a steady stress, which may be produced by thermal gradients, a solid can show the phenomenon of creep. It can plastically deform.

This inelastic response to stress is a well-known phenomenon in solid-state physics. Solids are never absolutely regular arrays of atoms or molecules. They contain atomic-sized holes called lattice vacancies, larger holes formed by clusters of vacancies, tunnels formed at the edge of an extra plane of atoms called edge dislocations, twists in the lattice called screw dislocations, arrays of edge dislocations at the boundaries of grains, and large mis-matches at the junctions of crystallites. Vacancies allow diffusion of atoms to take place, but dislocations are the principle sources of creep, since they can move relatively easily through the solid and carry their intrinsic deformation with them. Moreover the higher the temperature the faster they move. Solids are not entirely elastic. They have a reluctant plasticity about them which is analagous to a fluid with an extremely high viscosity.

This phenomenon suggests that convection cells may exist in the mantle. Hot material may be slowly rising towards the surface at certain points and cold material slowly sinking at others. Work by Rayleigh and Jeffreys, based on the assumption that the mantle was a highly viscous fluid, showed that such convection cells must have dimensions of the order of 2,000 km. Recent estimates of the rate of creep suggest that this becomes very small in the highly compressed lower mantle, so that convection cells are almost certainly limited to the upper mantle. If so, their shape is likely to be elongated horizontally, with short vertical upflows and downflows (Figure 4.6).

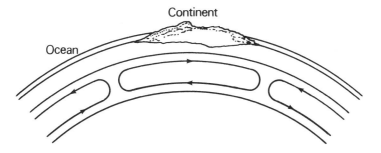

Figure 4.6 – Convection cells in the asthenosphere

Upper mantle convection provides a source of power for the observed drift of the continents, a topic we will discuss in a later chapter. If continental drift is taken to reflect the motion and size of a typical convection cell. we conclude that the convection velocity is a few centimetres every year and the radius of the cell is of order 2000 km. Evidence for the existence of a partially molten layer, the asthenosphere lying between the depths of 100 km and 250 km, is obtained partly from the observation of anomolously low seismic-wave velocities at these depths, and partly from calculations made on heat flow in the mantle. Convection within this layer would provide the force which drives the continents, a topic we shall discuss in Chapter 6. The asthenosphere is therefore a highly significant part of the earth's structure.

FURTHER READING

Bullen, K. E., 'The Interior of the Earth' (p. 56, *Scientific American*, March 1955).

Burke, K. C., 'Hot Spots on the Earth's Surface', (p. 46, *Scientific American,* August, 1976).

Cloud, P., ed., 'Adventures in Earth History', (W. H. Freeman and Co., San Francisco, 1970).

Emiliani, C., 'Pleistocene Temperatures', (*Journal of Geology,* **63**, 538, 1955).

Fraser, *Understanding the Earth,* (Penguin, 1967).

Gaskell, T. F., *Physics of the Earth,* (World of Science Library, 1970).

Gribbon, J., *The Climatic Threat,* (Fontana, 1978).

Holmes, A., *Principles of Physical Geology,* (Ronald Press Co., New York, 2nd edition, 1965).

Lomnitz, C. and L., 'Tangshan 1976: a case history in earthquake prediction', (*Nature,* **271**, 109, 1978).

Press, F., 'Earthquake Prediction', (p. 14, *Scientific American,* May, 1975).

Richter, F., 'Convection Currents in the Earth's Mantle', (p. 72, *Scientific American,* November 1976).

Siever, R., 'The Earth', (p. 82, *Scientific American,* September 1975).

Siever, R., 'The Steady State of the Earth's Crust, Atmosphere and Oceans', (p. 72, *Scientific American,* June, 1974).

Sylvester-Bradley, P. C., 'The Search for Protolife', (*Proc. R. Soc.,* London, **B 189**, 213 (1975)).

Wyllie, P. J., 'The Earth's Mantle', (p. 50, *Scientific American, March 1975).*

Wyllie, P. J., 'The Earth's Mantle', (p. 50, *Scientific American,* March 1975).

5

The Earth's Crust

Where the dust of an earthy today
Is the earth of a dusty tomorrow!

Sir W. S. Gilbert (1836–1911) *Patience*

5.1 ROCKS

The continents are essentially huge blocks of granite, with roots that reach
30 to 40 km below the surface to the denser mantle beneath. Inevitably a great
deal of continental structure is inaccessible to us, and we have to rely on information inferred from seismology. Only the comparatively thin surface layers
can be studied directly, but such a study has been the cause of a veritable
revolution in man's awareness of his own, and his planet's history.

The data are the rocks, exposed by mountainous upthrusts, or by the action
of wind and water, or by volcanic activity, or by deliberate drilling. There are
basically two sorts of rock — igneous and sedimentary (Figure 5.1). Igneous
rocks, like granite and basalt, have been thrust to the surface, often through
other formations, as a result of volcanic activities, and are glimpses of the underlying continent. Lava flows, from ancient volcanoes, forming layer upon layer of
once-molten material, and veins of minerals, threading their crystalline way
through the surrounding strata, give evidence of hot, violent activity. Sedimentary rocks, like limestone, chalk and sandstone, provide evidence, on the other
hand, of long, quiet periods of deposition in the ocean depths. Horizontal strata
of immense thickness, (such as the limestone crags of the Pennines, to give a
local example) point to unimaginable eons of time, in which the dendritus of
oceanic life and continental erosion, settled and became squashed by overlying
strata into hard, solid rock. The stratification itself is a graphic indication of
ancient history. The more modern layers overlying the more ancient provide a
time-sequence of conditions which suggests substantial variations of climatic
and other conditions. Fossil animals reflect the same variations, and more
importantly, exhibit the vastly long history of evolution. And what change is
suggested by the mere fact that such rocks are now exposed to the air on dry

land, that were once the ocean floor! Moreover there is abundant evidence that the reverse has occurred many times: weathered rocks buried under oceanic deposits. (Figure 5.1(c)).

The thickness and fossil record of sedimentary rocks have long been used to construct a geological time scale. Estimates of age have been made from the rate at which sediments accumulate, and the similarity between fossils found in various parts of the world has been used to establish contemporaneity of geographically separated strata. It was possible, using such methods, to establish an approximate time scale for the eras of sedimentary deposition from recent times to the beginning of the Cambrian period, 600 million years ago. Beyond that, apart from certain microscopic structures, fossils cease to exist. Yet this time span of 600 million years is rather small compared with, say, the age of oceans. Some indication of the latter is given by the rate at which salt is washed off the land into the sea to add to the briny tang. Holmes, the British geologist, estimated that the amount of salt in the sea today entailed a period of at least 2,000 million years. Sedimentary rocks of this age are therefore implied, but few can have avoided being metamorphosed by igneous activity. Nevertheless ancient rocks aged 2,000 million years and more have been discovered. One famous example is the gold-bearing conglomerate in South Africa.

Figure 5.1(a) *(below)* — Sedimentary rocks. Carboniferous limestone cliff of well-bedded limestones, shales and mudstones, near Castlemartin, Pembroke, Wales.
(Photo· T. C. Hall, 1909. Geological Museum, London)

10 centimetres

Fig. 5.1(b) – Volcanic rocks. *Above:* Basalt, Andesite, Granite. *Below:* Pumice, Lava, Trachyte, Obsidian.   (Geological Museum, London)

Above: Old Red Sandstone 375 My   Below: Lower Carboniferous 330 My

Figure 5.1(c)
Space views
of Britain
through
geological
time.
(Pages 66-68.
Geological
Museum,
London)

Above: Permian 250 My

Below: Jurassic 165 My

Above: Eocene 55 My

Below: Penultimate glaciation 250 ky

5.2 DATING BY RADIOACTIVITY

A much more accurate, quantitative, dating technique is to exploit the decay of naturally occurring, radioactive atoms. An element has characteristic chemical properties determined by the number and arrangement of the electrons which move around its nucleus, and the number and arrangements of electrons is determined by the number of protons in the nucleus. The latter is called the atomic number. The number of neutrons in the nucleus, together with the number of protons, determine the atomic weight. Atoms with the same atomic number, and therefore identical chemical properties, but with differing atomic weights are known as isotopes. Some isotopes are unstable, and spontaneously decay into different elements. The decay appears to be quite random, but with large populations, statistics can provide a definite measure of the average rate at which radioactivity takes place. One measure of this is the half-life, the average time it takes for half the number of radioactive atoms to decay. Isotopes used in radioactive-dating have half-lives of the order 10^9 years.

Radioactive atoms found in rocks do not, of course, begin to decay only when the rock was formed. They were decaying ever since they were produced in some supernova explosion, perhaps 10^{10} years ago. The dating method can work only if the rock captures and retains the stable daughter-elements of the decay process. Suppose an isotope X radioactively decays into a stable isotope Y of a different element. We can measure the amount of Y in a given sample of rock, and if we subtract from this the amount we would expect to find in the absence of radioactivity, we arrive at a measure of the amount of X which has decayed since the rock became solid enough to trap the daughter-element. A measurement of the amount of X present, and a knowledge of the half-life, then gives us the age of the rock.

The most important isotopes used in radioactive dating are the uranium isotopes U^{238} and U^{235}, thorium Th^{232}, rubidium Rb^{87}, and potassium K^{40}. The decay of uranium initiates a chain of radioactive decays which finally produces a stable isotope of lead. U^{238} decays to Pb^{206} with a half-life of 4.51×10^9 years, and U^{235} decays to Pb^{207} with a half-life of 7.13×10^8 years. In a given sample the concentrations of all components have to be measured. Th^{232} decays also via a series to a lead isotope, this time Pb^{208}, with a half-life of 1.4×10^{10} years. In contrast, the decay of rubidium is simple; the immediate daughter is a stable isotope of strontium Sr^{87}, the half-life being an immensely long 5×10^{10} years. Potassium behaves more awkwardly, 11% decaying to argon A^{40} and 89% decaying to calcium Ca^{40}, with a half-life overall of 1.3×10^9 years.

5.3 GEOLOGICAL ERAS

With these techniques the ages of the various geological periods back to the beginning of the developed-fossil record have been determined to an accuracy of a few percent. But now we can push much further back in time, and begin to produce a time map of the old Pre-Cambrian rocks which underlie the fossil-bearing strata. The oldest rocks to be discovered are in south-west Greenland. They are at least 3.76×10^9 years old. Rocks formed later than 2.6×10^9 years ago contain micro-fossils about 20 μm in size. Evidence of cell division (mitosis) approximately 1.5×10^9 years ago begins to provide a basis for the classification of Pre-Cambrian structures. We can now identify six eons of the earth's history, though the divisions are to some extent arbitrary.

The beginning, however, is not in the least arbitrary. The age of the earth itself has been determined by equating it with the age of meteorites, lumps of rock of extra-terrestrial origin, but belonging to the solar system, which have survived their incandescent journey through the atmosphere. Their radiometric age is 4.6×10^9 years, and this is believed to be the age of the solar system, and therefore of the earth. Rocks older than the geological structure at 3.76×10^9 years form the Eogeological Basement. From there to the appearance of proto-life at 2.6×10^9 years are the Archaean structures. Then follow the Proterozoic eons, divided into Early (2.6 to 2.0×10^9 years), Middle (2.0 to 1.0×10^9 years) and Late (1.0 to 0.6×10^9 years). Finally comes the Phanerozoic eon, containing the fossil-bearing rocks of the Cambrian period, and succeeding periods up to the present day (Table 5.1). The more detailed knowledge we have of this latter eon allows us to subdivide this into many periods familiar in geology (Table 5.2).

Table 5.1 – Earth eons

Eon	Time (Millions of years)	Biological record	Geological record
	– 0 –		
1. Phanerozoic		developed fossils	Mountain Building
	– 600 –		
2. Late Proterozoic			
			Mountain Building
	– 1,000 –	origin of sex	
3. Middle Proterozoic		mitosis (1500)	Mountain Building
	– 2,000 –		
4. Early Proterozoic			
	– 2,600 –	microfossils	Mountain Building
5. Archaean			
	– 3,760 –		
6. Eogeological Basement	– 4,600 –		

Table 5.2 – Phanerozoic Periods

Era	Period	Age $(\times 10^6$ yrs$)$	life	Mountains
Kainozoic	Quaternary			
	Recent			
	Pleistocene	$-$ 1 $-$		
	Tertiary		Man	
	Pliocene	$-$ 12 $-$		
	Miocene	$-$ 27 $-$		
	Oligocene	$-$ 42 $-$		
	Eocene	$-$ 60 $-$		Alps, Himalayas, Rockies
	Paleocene	$-$ 70 $-$		Pyrenees, Andes
Mesozoic	Cretaceous	$-134-$		
	Jurasic	$-175-$	Mammals + Birds	
	Triassic	$-220-$	Reptiles	
Paleozoic	Permian	$-260-$		Urals
	Carboniferous	$-330-$		Harz, Central Massife
	Devonian	$-380-$	Land animals	Appalachians, Caledonians
	Silurian	$-430-$		
	Ordovician	$-500-$		
	Cambrian	$-600-$	Invertebrates	

5.4 MOUNTAINS

Mountains are intensely folded structures, They are the result of immense upthrusts and compressions in the earth's crust, which have often folded, torn and shattered the quiet horizontal layers of sedimentary rocks, and interposed huge chunks of basal granite. No-one can doubt that their origin is an entirely violent one.

One might expect that such huge heaps of rock, piled up on the earth's surface to heights considerably above the mean elevation of land (840m above sea-level), would produce appreciable gravitational anomolies. A plumb-line in North India should certainly be deflected to the north by the Himalayas, but, in fact, this does not occur to anything like the extent one expects. The pile-up of material on the earth's surface is largely compensated by the same material reaching down into the dense mantle. Mountains appear to be rather like logs floating in water. The bigger the log, the more it sticks out of the water and the further it reaches below the surface. Mountains, containing rock of density about 2.6 gcm^{-3}, float on the denser (3.4gcm^{-3}) mantle.

This idea of floating is the basis of the theory of isostasy. Variations in the earth's crust are supported and balanced hydrostatically, with the denser acting as the 'liquid'. Mountains, and even continents and oceans, float on the upper surface of the mantle like so much debris on a pond. All vertical columns of crust with the same cross-sectional area have the same mass, so no large gravitational anomolies are to be expected.

The hot-cold daily cycle, and the wind and rain, erode away the sharp corners of mountains and cleave them with ravines and valleys (Figure 5.2). Young mountains, such as the Alps, the Himalayas, the Rockies and the Andes, still have their sharp edges — their peaks and ridges, and awe-inspiring precipices, yet unsoftened by erosion. Older mountain ranges, like the Scottish Highlands are gentler, more rounded, and more ancient ones still have to be inferred rather than directly perceived. There appear to have been five major mountain-building periods in the earth's history: three inferred from intense rock deformation and igneous intrusion aged around 2,600, 1800, and 1,000 million years ago, and two more recent episodes at about 250 and 70 million years ago. The older of the more recent ranges have very little in the way of earthquake and volcanic activity associated with them; the younger ones have a great deal. The processes which have formed these juvenile ranges are by no means quiescent, and we are reminded of this with every earth tremor and volcanic eruption.

Figure 5.2(a) *(below)* – Hadramaut, Saudi Arabia.   (Photo: NASA)

Figure 5.2(b) (*above*) – Hadramaut from aircraft.
(Photo: R. S. Scorer)
Figure 5.2(c) (*below*) – Gully erosion. Ashdown Forest, Sussex, England. (Geological Museum, London)

5.5. OCEANIC RIDGES

Mountain ranges are not confined to the continents. Rising from the ocean floor in the middle of the major oceans are immensely long ridges ascending some thousands of metres and occasionally surfacing as volcanic islands, such as the Azores in the mid-Atlantic (Figure 5.3). The mid-Atlantic ridge runs for thousands of kilometres right down the middle of the North and South Atlantic oceans, like a dividing wall between America to the west and Europe and Africa to the east, passing right through Iceland, to which it occasionally adds a volcanic island. Another ridge separates Antartica from the neighbouring continents to the north, a third runs north-south some 4,000 km to the west of South America, and a fourth divides the Indian Ocean (Figure 5.4).

Although they are elevated structures, oceanic ridges are not at all similar in structure to mountain ranges. They are not intensely folded structures, but vast mounds. Running down the centre is a parallel-sided valley, somewhat like the Rift Valley in East Africa. All the indications are that the compression which appears to be associated with mountain building is entirely absent, and that the driving mechanical force is a pulling-apart rather than a squeezing together. A network of transform faults in the ridges points to the effects of gigantic stretching and shearing forces. These transform faults are where the crust has fractured as regions have been displaced horizontally relative to their neighbours. Such fracture zones cross the ridge at right-angles in straight lines which may sometimes be over a thousand kilometres long, and they can give rise to some impressive undersea features. For example, the 2,000 km long Mendocino fault stretching westwards from California has a south-facing cliff 2,000m to 3,000m high; evidence of a small vertical component of shear in the vast forces which have produced a relative horizontal displacement of some 1,000 km! Not surprisingly, transform faults, like the younger mountain ranges, are zones of intense earthquake and volcanic activity (Figures 5.3 and 5.4).

5.6 OCEANIC TRENCHES AND ISLAND ARCS

The mean depth of the ocean is about 3,800m, but there exist long valleys much deeper than this, known as oceanic trenches. Stretching for hundreds, and often thousands of kilometres, almost 100 km wide, these trenches reach depths of 6,000 to 10,000m. The major valleys run along the north and west boundaries of the Pacific Ocean (Figure 5.4). A trench line can be traced from Alaska, which arcs south of the Aleutian Islands towards the Kamchatka Peninsular of Siberia, turns south following the Kuril Islands towards Japan, and then curls off towards the Mariana Islands. Another skirts the east coasts of Japan and the Philippine Islands, and one stretches south-east from New Guinea towards the Kermadec and Tonga trenches running north-east from New-Zealand.

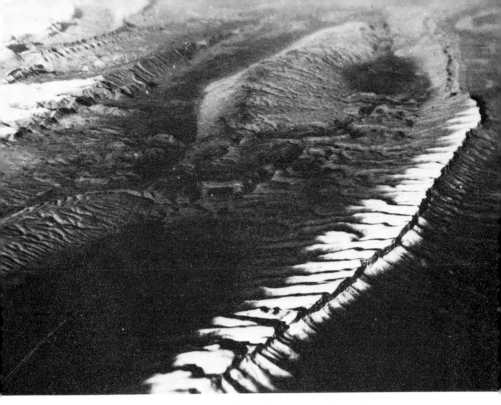

Figure 5.3 – Geological features. (a) *(above)* Snow-capped ridge in Iranian desert west of Teheran. (Photo: R. S. Scorer)
(b) *(below)* Fault (Photo: Geological Museum, London)

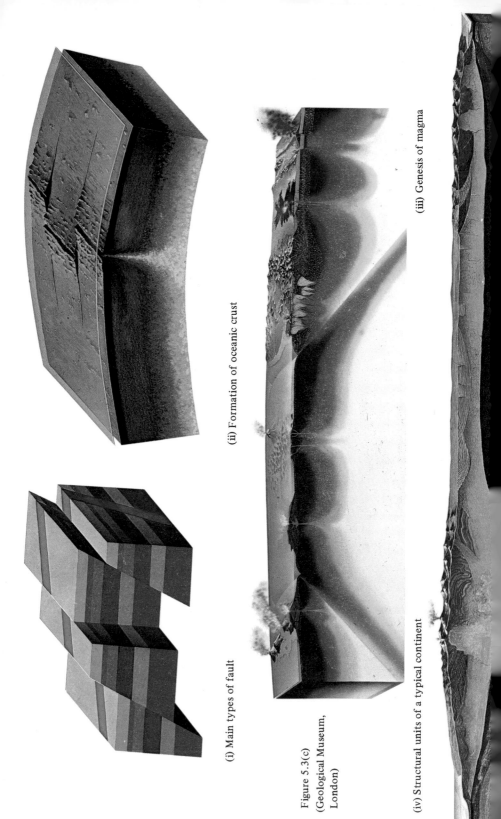

(ii) Formation of oceanic crust

(i) Main types of fault

Figure 5.3(c)
(Geological Museum,
London)

(iii) Genesis of magma

(iv) Structural units of a typical continent

It is striking that these trenches are found to run parallel with lines of volcanic islands. These island arcs are a characteristic feature of the Pacific Ocean. Examples are the Aleutians, the Kurils, the Japanese Islands and the Marianas. All are notorious for their earthquake and volcanic activity. In this curious pairing of oceanic trenches and island arcs, we have another major feature of the earth's crust, which manifests a crustal instability as potent as the oceanic ridges and continental mountains.

5.7 EARTHQUAKES

Earthquakes and volcanoes, far from being found uniformly spread over the surface of the earth, occur in fairly well-defined zones (Figure 5.4), almost always associated with one of the features of the crust we have just been discussing. These zones are where the earth is active. Immense forces are at play which implaccably build up mechanical and thermal stresses which can be temporarily relieved only by earthquakes and volcanic eruption. Such activity is erratic but every time it occurs life, in these zones, is at risk. The physical environment in such zones can be deadly.

One of the most urgent needs in highly populated regions like Japan, China and California, where the risk is high, is for a reliable means of predicting earthquakes. Volcanoes can be given a wide berth, but earthquakes cannot be avoided. There is no way in which we can defend a city from being destroyed by a large earth tremor — the forces involved are too great for the puny technology of our day to influence — but we can at least save its citizens if we can predict such a catastrophe. We may hope that ways will be found to release the stress in a controlled manner and thereby to avoid the more violent earthquakes, but until we can do that the only defence is evacuation. And this is possible only if we can predict. Moreover the prediction has to have a high probability of being accurate. The evacuation of a large city is not to be undertaken lightly — the social disruption would be considerable.

There are many small earthquakes and very few large ones in a given period of time. Earthquake intensities vary over so many orders of magnitude that it is convenient to measure on a logarithmic scale rather than a linear one. The Richter scale is essentially based upon the ratio of the amplitude of ground motion a, in microns, to the period T of the dominant wave in seconds. The magnitude M is approximately given by

$$M = \log_{10}\left(\frac{a}{T}\right) \tag{5.1}$$

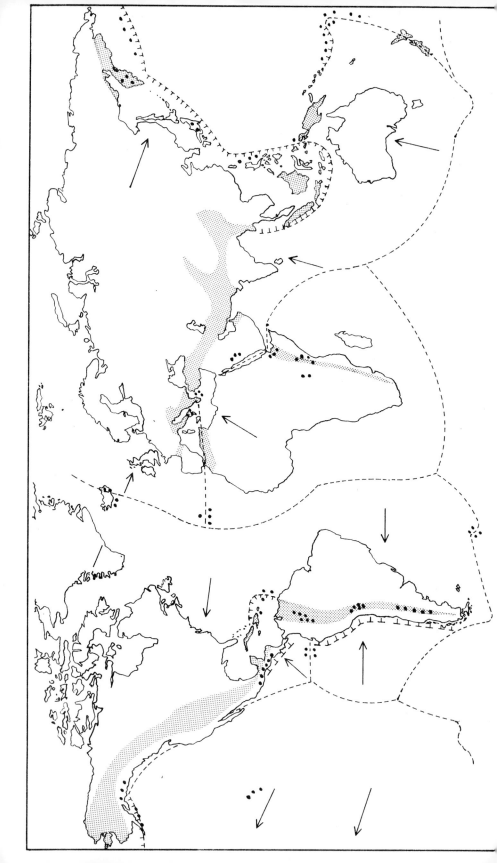

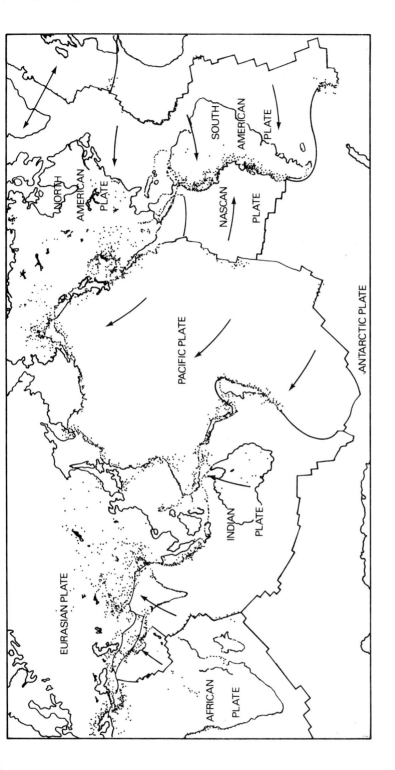

Figure 5.4(b) – Epicentres of earthquakes between 1961 and 1967.

(A more accurate expression involves corrections for local conditions and distance from the source of the earthquake). Towards the lower end of the scale a high frequency tremor of about 1,000Hz with an amplitude of as little as a few microns, so that M is about 3.5, can be just detected when one is sitting or standing. Towards the higher end surface waves some metres in amplitude ($a \approx 10^7 \mu m$) with a period of about 10s, giving a magnitude of about 6, can produce terrifying effects. The largest earthquake recorded in recent times had a Richter magnitude of no less than 8.7. It occurred in Assam in 1952. Another in Alaska in 1964 had a magnitude of 8.4. The disastrous Tangshan earthquake of 1976 in China had a similarly high intensity, 7.8. At the other extreme sensitive seismometers can record local shocks down to a magnitude −3. Damage becomes serious at about magnitude 5.

Fortunately there are several clues which point towards a forthcoming large earthquake. Some of these, which are at present actively investigated, are:

1) frequently, but not always, a lull in the occurrence of small earthquakes in the preceding month.
2) a reduction in the ratio of P-wave velocity to S-wave velocity;
3) anomalous changes in the volume of crustal rock, as recorded by tilt-meters, changes in sea-level, and repeated surveying;
4) changes in water level, water turbidity and temperature, and radon content of deep wells;
5) an abrupt decrease in electrical resistivity of the rocks;
6) changes in magnetic field;
7) unusual behaviour of animals.

But so far only about ten earthquakes have been successfully predicted. What causes these changes is still controversial. When a rock is subject to intense pressure it deforms and eventually crumbles, but before breaking it develops tiny cracks which causes it to dilate, and this changes its properties in a significant and, in principle, detectable way. The details are insufficiently understood, but hopes are high that this phenomenon of dilatancy will ultimately provide the physical basis for the accurate prediction of large earthquakes.

At present the prediction of serious earthquakes is bedevilled by the variations of quantitative as well as qualitative forerunners which occur. Nowhere was this more tragically illustrated than in China. On 29th May 1976 in Southern Yunnan, the Lungling earthquake was successfully predicted about 20min before the main shock. In fact it turned out to be a twin event, with two shocks occurring at magnitudes of 7.5 and 7.6. Evacuation was just in time, and no lives were lost. On the 7th November and the 13th December 1976 the Yenyuan earthquakes on the Yunnan-Sze-Chuan border, magnitudes 6.9 and 6.8, were also successfully predicted 3 days in advance, and again no lives were lost. In between these successful predictions came the Tangshan event on 28th July, which consisted of a main shock of magnitude 7.8 followed by twelve aftershocks exceed-

ing magnitude 6. None of the warning signs successfully used to predict the
Lungling and the Yenyuan events were present. The result was a major disaster.

5.8 ICE AGES

There is abundant evidence on the earth's surface that there have been
periodic invasions by ice from the polar caps. Geologically speaking, the time
scale involved is short. The last ice-age, which covered much of Britain, Northern
Europe, Northern Siberia and North America with ice, occurred 20,000 years
ago. It left a legacy of debris carried south by advancing glaciers, gouged-out
valleys channelling tiny streams as inadequate reminders of the ice which once
filled them, coombes, or corries, in the hills which once exuded huge rivers of
ice, and large stretches of inland water such as the Great Lakes of North America.
In the past 300,000 years there have been seven major ice-ages, with warm
periods and minor ice-ages in between. Roughly every 40,000 years the earth,
in recent times, has had to endure a period, which may last for a thousand years,
in which the temperate zones become arctic. Cool spells, of various degrees
of severity, occur roughly every 13,000 years, during the 'warm' periods. It is
possible that we are moving into such a cool spell at present. (Figure 5.5).

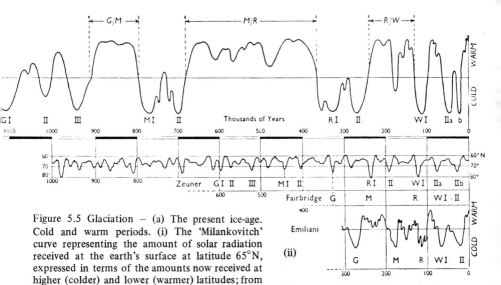

Figure 5.5 Glaciation – (a) The present ice-age.
Cold and warm periods. (i) The 'Milankovitch'
curve representing the amount of solar radiation
received at the earth's surface at latitude 65°N,
expressed in terms of the amounts now received at
higher (colder) and lower (warmer) latitudes; from
data recalculated by A. J. J. van Woerkom for the last million years. Correlations
favoured by F. E. Zeuner and R. W. Fairbridge are indicated. (*Curve after C.
Emiliani,* 1955, Fig. 14). (ii) C. Emiliani's correlation between the 'Milankovitch'
curve and a generalised climatic variation curve based on deep-sea cores (*After C.
Emiliani,* 1955, Fig. 15). (Arthur Holmes, *Principles of Physical Geology,* 1965.
Ronald Press Co., N.Y. by arrangement with Thomas Nelson & Sons Ltd, Sunbury,
England)

Figure 5.5(b) (*above*) – Glacier near Mt. Blanc. (Chamonix is to the left). (Photo from aircraft, R. S. Scorer)
Figure 5.5(c) (*below*) – Penultimate glaciation of Britain 250 ky ago.   (Geological Museum, London)

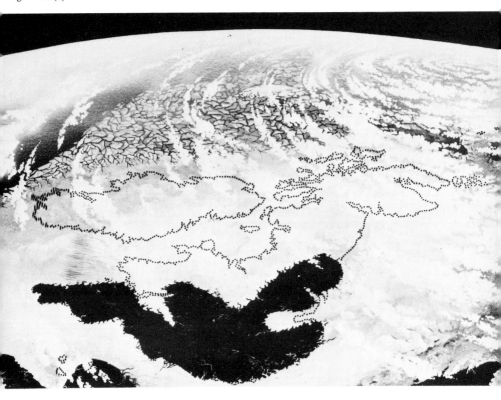

   Ice ages are not confined to one hemisphere or the other. They appear to be associated with world-wide changes on climate. Perhaps the most striking evidence for this is provided by the analysis of oxygen isotopes in the calcium carbonate $CaCO_3$, of fossil shells. The natural abundance ratio O^{18}/O^{16} is about 1/500, but the isotope O^{18} is slightly more concentrated in $CaCO_3$ than in water, H_2O. Moreover the excess is greater the lower the temperature, so a careful analysis can reveal the temperature of the ocean when the organism was alive. The temperatures deduced in this way from tropical, deep-sea cores show fluctuations between $22°C$ and $29°C$ over the last 300 thousand years, which correspond with the geologically determined dates for the ice ages of the recent Pleistocene period. An ice age is therefore associated with a global drop in temperature.

   The cause of such a drop in temperature is still highly contentious. The factors which enter are as follows:

(1) Solar luminosity

   The most direct cause would be a slight drop of solar radiation. The sun is a gigantic hydrogen bomb, maintained in a stable form by its own gravitation. It is possible that slight fluctuations in output may occur, and indeed a variation by about 0.5% in the amount of sunlight received by the earth has been observed this century. Such fluctuations are likely to be an intrinsic property of the sun, but large changes could occur if the solar system, in its journey through the outer reaches of the galaxy, passed through a cloud of gas or dust.

(2) Orbital characteristics

   The amount of solar radiation falling on the earth's hemispheres varies cyclically as the inclination of the earth's axis changes, as the axis precesses, and as the orbit itself undergoes slow variations. The angle which the rotation axis makes with the normal to the orbital plane (the obliquity of the ecliptic) varies between $24.60°$ and $21.98°$ and completes the cycle in 40,000 years. The precession of the equinoxes has a period of 25,800 years, and the orbit changes its shape between an ellipse and a circle with a period of 92,000 years. These periodicities are of the right order of magnitude to fit the observed periodicities of ice-ages. However, since none of these variations changes the mean distance between earth and sun, they do not change the average amount of solar radiation falling on the earth as a whole, so in themselves they cannot account for simultaneous glaciation in both hemispheres. Nevertheless, the two hemispheres are not symmetrical, and orbital variations do induce small climatic changes, so their effect cannot be ruled out.

(3) Cooling of the Oceans.

   The oxygen-isotope thermometer, applied to fossils dating back to Jurassic times, shows that the oceans have been cooling over the past 50 million years.

The inference drawn from this and other geological data, is that the global climate was significatly warmer than it is now, during the Late Mesozoic and Early Kainozoic times. There is no evidence of glaciation until we reach the Permian and Carboniferous periods in Gondwanaland 200–300 million years ago. This suggests that long-term changes in climate, with a period of some 250 million years, underlie the recent fluctuations responsible for the Pleistocene ice-ages. Such long-term effects may be associated with thermal cycles within the sun.

(4) Albedo

The percentage of solar radiation reflected back into space is known as the albedo. Land has an albedo of about 20%, calm sea 8%, stormy sea 40%, clouds 70% and ice 70%. The high albedos of clouds and ice make the area of the globe covered by these entities an important factor in determining terrestrial temperatures. Clouds are probably not an important factor in the triggering of an ice-age. In general, clouds form on upcurrents of air and clear skies occur on downcurrents. Since at any moment there must be as much air rising as falling over the surface of the earth, the cloud cover is expected to be about 50%. This indeed, is what is observed, although continents generally have clearer skies than oceans. There seems little reason to suppose that this 50% cover changes with time in an appreciable way, and so it is unlikely that an ice-age is triggered by a sudden increase in cloudiness. Indeed, it is far likelier that clouds are stabilising influences on climate. Less heating means less evaporation and fewer clouds, and fewer clouds means less reflection of sunlight and more heating. This mechanism provides a potent negative feed-back which tends to maintain equilibrium. In the case of ice and snow, however, there is the possibility of a positive feedback mechanism which may drive an instability. More ice and snow means more reflection and hence more cooling. This may well be an important factor during glaciation.

(5) The Greenhouse effect

The atmosphere is largely transparent to visible light, which is the part of the spectrum containing the peak solar energy. The earth, warmed by solar light, radiates a spectrum whose peak is in the infrared at a wavelength of about 10μm. This radiation cannot escape immediately because it is absorbed by the atmosphere, particularly by water vapour and to a lesser extent by carbon dioxide (CO_2). The atmosphere therefore acts in the same way as the glass in a greenhouse, letting through light but not allowing infrared radiation to escape. (Figure 5.6). It helps to retain the earth's heat. (Actually, greenhouses are warm mainly because air is not allowed to escape and convect heat away – a crop of standing corn, a walled garden, a tree-bordered enclosure, act in the same way and become warmer than their surroundings by reducing the free circulation of air. To some extent, therefore, 'the greenhouse effect' is a misnomer!) Any reduction in the water vapour or CO_2 content of the atmosphere would weaken the green-house effect and allow the earth to cool. Like the albedo mechanism there is the possibility here of positive-feedback. More water tied up as ice means less water vapour to act as a radiation blanket. But first the water has to be tied up as ice. . . . . . .

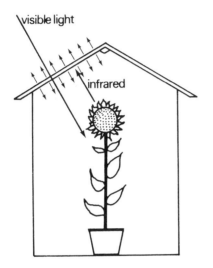

Figure 5.6 – The greenhouse effect. Glass is transparent to visible light but absorbs infrared light.

(6) Dust

Recent observations by satellite show that the high atmosphere contains a zone of dust, which may have originated from meteors. In a day the earth picks up about a million tons of solar system debris, and it is possible that this process has an influence on climate, the dust preventing the sun-light reaching the ground. It is conceivable that an ice-age could be triggered off by the passage of the earth through a huge cloud of meteors. Another source of dust in the atmos-phere is volcanic activity, and this raises the possibility of a connection between climatic changes and periods of intense vulcanism. A link with cloud formation is provided by the fact that dust acts as nuclei for condensation. However too little is known about the possible variations in dust concentration for a connection to be made with the initiation of an ice age.

(7) Distribution of continents

The position of continents on the globe has by no means remained constant, as we shall see in Chapter 6. The present distribution of land masses is such that a continent lies astride the South Pole; and the North Pole, though oceanic, is almost land-locked. Both of these features make it impossible for enough warm sea from the tropics to reach the poles and inhibit ice-formation. The present-day polar ice-caps are therefore a direct result of continental distribution, and in earlier geological ages would not have existed for a great part of the time. Our present epoch is therefore peculiarly prone to glaciation.

5.8.1 What causes ice-ages?

What causes ice-ages is a question of immense interest. Are we moving into one presently? How fast does an ice-age happen? Can anything be done to prevent it? The fact is we know too little to answer any of these questions.

The factors we have been considering make the origin of ice-ages exceedingly difficult to pin down. Many different physical processes enter and any or all could contribute to some degree. The long gap between the Permo-Carboniferous glaciation and the recent (one should say present) ice-age era suggests that the slow variations in solar luminosity may be responsible at base, but the cause of the relatively rapid fluctuations in global temperature which have accompanied the various periods of the Pleistocene glaciations is by no means understood. The slow cooling over the past 50 million years has eventually primed the earth for an era of ice-ages, but what triggers off a bout of glaciation is still a lively topic of debate.

Even the long period between the ice-age eras may not be a constant. Before the Permo-Carboniferous glaciation there is evidence for a Late Pre-Cambrian ice-age era, which fits with a rough 250 to 300 million year cycle, but earlier periods of glaciation do not appear to occur so regularly. Whether such irregularities can be explained in terms of an irregular solar output is a problem for astrophysics to solve. It may be, however, that the periods between ice ages is as little as 150 million years. The Australian physicist Williams identifies a cool period around 150 million years ago, the Permo-Carboniferous glaciation 295 million years ago, a late Ordovician glaciation 445 million years ago, and three in the Late Precambrian at 615, 770 and 940 million years ago. He relates this 150 million year period to half the cosmic year – the time for the solar system to orbit the galaxy. This would suggest a truly cosmic origin for ice-ages. Twice a galactic year the solar system perhaps passes through a dust cloud and this may trigger off an ice-age. On the other hand, the history of continental drift reveals a period of about 400 million years connected with continental disruption, scattering and reunion. Ice-age eras may coincide with the irregular occurences of polar continents.

Very many physical factors enter into the problem, and it is probably true to say that, intrinsic solar fluctuations apart, no one factor on its own is adequate to explain the onset of glaciation. The most basic is the amount of radiation falling on the earth, and it is possible to estimate how this varies as the orbital characteristics of the earth vary, by assuming that the sun's output remains the same. The most notable results of such a calculation for the last

600,000 years, were obtained by the Yugoslavian physicist M. Milankovitch in 1930, who discovered long periods of cool summers alternating with periods of warm summers. He speculated that long periods of cool summers would allow snow which fell in the winter to remain unmelted, and that this would trigger off permanent snowfields and glaciation, and indeed a good case can be made for corelating these orbital cool spells with the geologically determined glacial phases. (Figure 5.5).

5.9 MINERALS AND FOSSIL FUELS

Our ability to mould the physical world into structures that we find useful, such as moon-rockets, computers, cars, dishwashers, guns, kettles, spoons and safety pins, depends upon our skill at manipulating the very rocks of the earth's crust. These rocks are our technological, as well as our physical base. What these rocks contain and consist of determines our freedom of action more potently in the long run than anything else.

That being so, it is daunting to discover that of the ninety or so stable elements which the universe contains, only a tenth of that number appear in any quantity in the rocks. In fact, as Table 5.3 shows, a mere dozen elements account for 99.2% by weight of the earth's crust. Indeed, eight elements alone account for over 98%. In the list of elements given in Table 5.3 one may note the absence of useful metals such as copper, tin, lead, gold, silver, platinum, chromium, tungsten and uranium.

Table 5.3 — Percentage composition by weight of the Elements in the Earth and in The Earth's Crust

Earth's Crust		Whole Earth	
1. Oxygen	46.6	1. Iron	~45
2. Silicon	27.7	2. Oxygen	~25
3. Aluminium	8.1	3. Silicon	~12
4. Iron	5.0	4. Magnesium	~9
5. Calcium	3.6	5. Nickel	~5
6. Sodium	2.8	6. Calcium	~2
7. Potassium	2.6	7. Aluminium	~1
8. Magnesium	2.1		————
9. Titanium	0.4		99
10. Hydrogen	0.1		————
11. Phosphorous	0.1		
12. Manganese	0.1		
	————		
	99.2		
	————		
13. Sulphur	0.05		
14. Carbon	0.03		

Table 5.3 was extended to include carbon, the base element of organic life. All but a tiny fraction of the carbon in the crust exists in the form of carbonates of one sort or another in limestone deposits. Much of that element exists as carbon dioxide in the atmosphere, taking part in a highly complex cycle involving living things. Of all the organic matter which has been alive on the Earth's surface, a tiny percentage has become buried or trapped in sediments of one sort or another, and has failed to decompose through lack of oxygen. Wherever this has occurred we find a deposit of fossil fuel: peat, coal, petroleum, gas. Peat and coal originate from compressed vegetable matter, petroleum and gas from organic matter on the sea floor under the action of anaerobic bacteria. In France, Spain and Britain the coal fields date from the Carboniferous or later Cretaceous or Tertiary periods. Some oil fields are Pre-Carboniferous i.e. before the proliferation of land plants. The fact is that our fossil fuels, on which so much of our civilisation depends, are products of processes lasting up to 600 million years. Once used, there is no replacing them. Metals from minerals can, in principle, be used over and over again. Energy can be obtained from a fuel only once.

Table 5.4 shows the percentage composition by weight of elements in the oceans and in water-vapour-free atmosphere. Hydrogen appears abundant in water in the oceans but only as a trace as methane (CH_4) in the atmosphere.

Table 5.4 – Percentage composition by weight of Elements in the Ocean and Dry Air

Ocean			Dry Air		
1.	Oxygen	85.8	1.	Nitrogen	75.5
2.	Hydrogen	10.7	2.	Oxygen	23.1
3.	Chlorine	1.90	3.	Argon	1.29
4.	Sodium	1.06	4.	Carbon	0.01
5.	Magnesium	0.13			————
6.	Sulphur	0.09			99.90
7.	Calcium	0.04			————
8.	Potassium	0.04		Helium	0.000072
		————		Hydrogen	0.00002
		99.76		(in methane)	
		————			

Weight of earth = 5.977×10^{24}kg
Weight of oceans = 1.42×10^{21}kg
Weight of atmosphere = 5.30×10^{18}kg
Weight of crust* = 1.34×10^{22}kg

*Crust defined as spherical shell of width 10km, radius 6360km, density 2.64gcm⁻³.

Chlorine is present in the crust at a level of 0.03% but it is much more abundant in sea-water. The inert gas argon is present in relatively large concentration in the atmosphere. To get these figures in perspective, Table 5.4 also includes the weights of earth, crust, oceans and atmosphere. To a first approximation we may regard our mineral resources to be simply those of the crust, though this simple picture will be a poor one for elements such as hydrogen, chlorine and the inert gases.

Heavy materials will tend to sink downwards towards the centre of the earth, and this accounts for differences between the compositions of the earth as a whole, and the crust. The broad pattern of abundance is, however, not peculiar to the planet. Taking into account the loss of hydrogen and helium by the earth, we find the same pattern in meteorites and in the spectra of stars. At base, it is determined by the nuclear reactions with occur in stars as they evolve. Gold is scarce not only because it has sunk with the iron to the centre of the earth, but also because it has never been manufactured copiously by the stellar factories. The top eight elements are among the most abundant anywhere in the universe. Another factor which has governed terrestrial abundance is the position of the earth in the solar system. The inner planets, Mercury, Venus, Earth, Mars, are significantly denser then the massive planets Jupiter, Saturn, Uranus and Neptune. In the case of Jupiter and Saturn this is largely because their gravity is strong enough to retain the light elements hydrogen and helium, but in the case of the smaller, but still large planets, Uranus and Neptune, it is because their principle components are water, methane, ammonia and maybe neon, rather then the rock-iron mixture of the inner planets. The less volatile rocky substances have condensed out as the inner planets, and the more volatile gaseous materials have eventually condensed far from the sun. We have an inheritance which obtained its character in the very origin of the solar system, a topic we will return to in Chapter 10.

Over 70% of the crust consists of the elements oxygen and silicon. Thus, even in the case of the common metals, iron and aluminium, it is necessary to look for particular places where chance and geological activity have combined to concentrate metals into workable ores. Mineral deposits where such a concentration has occurred are therefore of immense value, and are continually being sought. Usually they are associated with ancient igneous activity where hot magma has been forced up from below into the overlying rock, and has solidified into veins and bands. Recently it has been discovered that circles of mineral deposits occur, apparently associated with old meteorite craters, suggesting that they have been formed by collision with extra-terrestrial objects.

Further Reading
   See pages 61 and 62.

6

Continental Drift

Sink down ye mountains, and ye valleys rise . . . . .
Be smooth ye rocks, ye rapid floods give way!
Alexander Pope (1688-1744) *Messiah*

6.1 EVIDENCE

A cursory glance at a map of the world shows land masses of all shapes and sizes in an apparently haphazard pattern. The suspicion that the pattern was not totally random dates as far back as the sixteenth century when Francis Bacon (1561-1626), perhaps the first philosopher of science, commented on the fit of the coastlines of South America and Africa, were these two continents to be pulled together. This coincidence continued to strike scientists and explorers in the nineteenth century, and in 1858 Antonio Snider went so far as to suggest that the shape of the coastlines was no coincidence, but caused by continental disruption. A crystallisation of such ideas came in 1910 when Alfred Wegener (1880-1931) proposed that the present form of the continents had come about by the disruption of one gigantic continent which he called Pangaea. The continents formed by the disruption had then drifted apart.

This was a revolutionary idea which, needless to say, struck a great many scientists of the day as the most ridiculous nonsense. There is Sir Lawrence Bragg's apocryphal story of the geologist who, on reading an article by Wegener commissioned for a British Journal, provided the only occasion in Bragg's life on which he saw a man literally foam at the mouth. Yet, at the present time, the evidence which has been collected over recent years is overwhelmingly in favour of the idea that continents do indeed drift.

The evidence may be summarised as follows;

(1) Shape

We have already mentioned the suggestive contours of the Atlantic coastlines of South America and Africa, but if one tries to fit other continents together on the basis of coastline contour, the fit is scarcely convincing. However an extremely convincing solution to the global jig-saw puzzle is obtained if the contours, not of coastlines, but of continental shelves are taken. Since a continental shelf is properly part of the continent it is therefore the more appropriate criterion for continental shape. That the jig-saw can be solved in this way is convincing evidence of the disruption of one or more super-continents. (Figure 6.1).

(2) Rock formations

The study of contemporaneous strata and general geological features strongly suggests the previous existence of two super-continents, which have been called Laurasia and Gondwanaland. Lines of folding which can be followed along the east side of North America to Scandinavia, show that North America was at one time joined to Europe. North America, Europe, Russia and Siberia once formed a vast continent which we call Laurasia. Strata of 2,000 million year old rock can be traced from the Atlas Mountains in North Africa right across to Venezuela in South America; rocks in Madagascar link with those in India; those in Australia suggest a link with Antarctica. Gondwanaland consisted of South America, Africa, India, Australia and Antarctica. The study of fossils, the flora and the fauna also suggest that there were two-super-continents in Mesozoic times, about 190 million years ago, and that these had largely broken up by the Cretaceous period, 70 million years ago.

Figure 6.1 – General pattern of continental drift. Top left, 380My; top right, 200My; bottom left 135My; centre, present; bottom right, 50My ahead.

(Geological Museum, London)

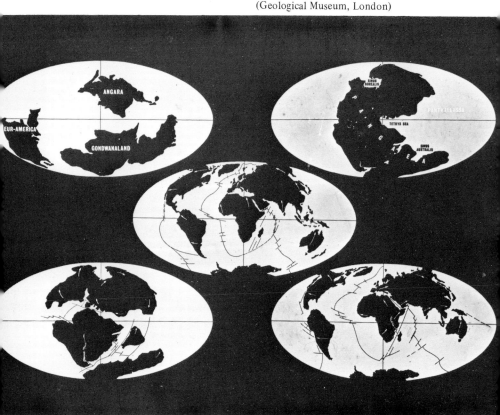

(3) Ancient climates

One of the most striking puzzles which geology throws up is the enormous variation of climate suggested by the different types of sedimentary rock found in a given geographical area. The British coal measures are records of a once tropical climate. The oil-fields of the North Sea, Canada and Alaska point to a warmer history than their present positions suggest. On the other hand, evidence of icy conditions recorded in one part of the world contemporaneous with the record of tropical conditions in another, preclude an explanation based upon major climatic changes over the whole globe. The record of marked variations of ancient climate therefore points towards continental drift. When the coal measures were laid down 300 million years ago, Britain was much nearer the equator.

(4) Paleomagnetism

Perhaps the most convincing evidence comes from the study of fossil magnetism. Besides showing that the earth's magnetic field has reversed frequently in the past (Chapter 4), the frozen record of the direction of the field relative to the horizontal, the angle of dip in other words, provides a good indication of the latitude of the rock when the magnetism was frozen in. A small angle of dip would suggest a position on the earth's surface near the equator, while a large angle of dip would point to an origin near one of the poles. The determination of latitude in this way can, however, be approximate only, since the magnetic pole wanders around the geographic pole. But, because we are interested in detecting processes extending over many millions of years, it is permissable to regard the earth's field, on average, to be that of a magnet aligned along the rotation axis, and so the average angle of dip in a rock layer tells us the ancient latitude. Studies of contemporaneous paleomagnetism within a continent, and in different continents, show incontrovertably that over some 700 million years the continents have wandered about the surface of the globe at speeds of a few metres a century.

(5) Sea-floor spreading

Equally convincing evidence is provided by the paleomagnetism of sea-bed rocks in the vicinity of the Mid-Atlantic ridge. A striking pattern of magnetic field reversals emerges (Figure 6.2). Immediately adjacent to the ridge on both sides run strips of rock showing the same direction of magnetism. Then follow adjacent strips showing reversed magnetism and these are, in turn, bounded by strips in which the magnetism has reverted to its 'normal' direction. The pattern of parallel strips of alternate directions of magnetization is repeated as one goes away from the ridge. Moreover the rocks are older the further they are from the ridge. This suggests strongly that the sea-floor is spreading outwards on either side of the ridge, so that the older rocks which once were close to the axis of the ridge have been displaced by younger rocks welling up from below to fill the gap. The magnetization of hot lava welling upwards and spreading outwards will be zero, but as the rock cools the magnetic particles it contains will be magnetized by the prevailing magnetic field of the earth and a record of the field direction will be frozen in. The study of this record gives us strong evidence of magnetic-field reversals, of sea-floor spreading, and its rate. The rate of spreading is about 4cm every year. This effect is to be expected if the continents on both sides of the Atlantic were moving apart. In the East Pacific, the rate is 8cm every year.

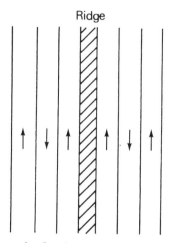

Figure 6.2 — Reversals of rock magnetism in strips running parallel to Mid-Atlantic Ridge.

6.2 PLATE TECTONICS

The principle features of Wegener's theory of continental drift appear to be confirmed rather well. The theory has, however, not remained static. It has evolved into the theory of plate tectonics, an evolution brought about by the appreciation that a continent was not the vital element involved, but rather the whole section of the solid lithosphere of which it was a part. The division of the crust into continents and oceans, was, in terms of structure, entirely misleading. The fundamental element is, in fact, a crustal plate, which may bear both continent and ocean. The whole crust can be subdivided into such plates, and it is the motion of these plates, driven by convection currents in the asthenosphere, which causes continents to drift and seas to spread.

But if the whole surface of the globe consists of moving plates, there must be a continual interaction between them at their boundaries. The triumph of plate tectonics is that it explains the formation of mountains, mid-oceanic ridges, oceanic trenches and island arcs, and the associated earthquake and volcanic zones. Wherever these features are observed on the surface of the earth,

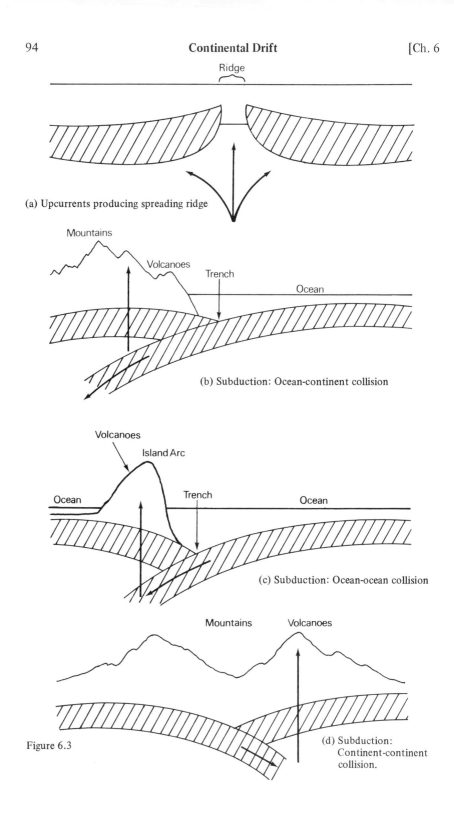

Ridge

(a) Upcurrents producing spreading ridge

Mountains

Volcanoes

Trench

Ocean

(b) Subduction: Ocean-continent collision

Volcanoes

Island Arc

Ocean

Trench

Ocean

(c) Subduction: Ocean-ocean collision

Mountains

Volcanoes

Figure 6.3

(d) Subduction:
Continent-continent
collision.

there are two or more plates colliding or separating. The earthquake and volcanic zones map out the plate boundaries. (Figures 5.3(c) and 5.4).

The basic idea is that the plate boundaries are situated wherever convection in the mantle produces up-currents or down-currents. Up-currents bring hot material to the surface, and in spreading out on either side, cause the plates to split and move apart. Mid-oceanic ridges are formed and sea-floor spreading is observed (Figure 6.3a). If the up-welling occurs beneath a continent, the latter will tend to be split apart. Such a process appears to be going on in East Africa up through the Red Sea in the Great Rift Valley. Part of Africa and all of Arabia appear to be twisting away from the main African continent. This particular convection current forced India into Asia some sixty million years ago, raising the Himalayas.

Down-currents cause plates to collide, and one plate is subducted below the other (Figures 6.3b.c.d). Light material of the crust is thrust deep into the hot asthenosphere. It melts and forces its way back to the surface in volcanic activity. The plate above is crumpled into mountain ranges in a continental collision, but in a collision of oceans, a pattern of deep trench and island arc develops. In all cases, plate boundaries are foci for extensive volcanic and earthquake activity, and the implaccable movement of one plate relative to the other, at the rate of a few centimetres a year, is the source of those striking transform faults, in which a large piece of land, or sea-bed, is shifted sideways relative to its neighbours. The San Andreas Fault in California is an example in which the mismatch of surface features along its length is particularly striking.

Before describing the present-day pattern of plates, it should be pointed out that sideways movement is not the only sort. Like rafts floating at sea, continents will sink lower as they become heavily loaded. Such a loading occurred during the last ice-age, when huge quantities of water were transferred from the oceans and deposited on the northern parts of America and Eurasia as ice. Since the ice melted and the glaciers retreated those parts affected are rising at rates reaching the order of a metre per century (e.g. Scandinavia, particularly the northernmost reaches of the Baltic Sea), in an attempt to establish isostatic equilibrium.

6.3 GLOBAL PLATES

Eight major plates have been identified (Figures 5.4 and 6.4). They are:

(1) Eurasian
Principally continental, it stretches from the mid-North-Atlantic ridge eastwards to include Europe, Russia and Siberia, and it is bounded to the south by the Mediterranean Sea and the Himalayas. It is presently travelling roughly south-eastwards.

(2) North American

Bounded to the west by the mountainous Pacific coastlines of Alaska, Canada, U.S.A. and Mexico, and to the east by the mid-North-Atlantic ridge, it is moving roughly north-westwards. The North Atlantic is growing steadily at a rate of about 4cm a year, and has reached its present width in some 120 million years.

(3) African

The plate is pushing roughly north-eastwards against Southern Europe and moving away from the Mid-South-Atlantic Ridge.

(4) Indo-Australian

A sprawling plate, stretching from the spreading line of the Red Sea — Indian Ocean in the west and colliding with the Pacific Plate in the east, it includes Arabia, India and Australia. Moving rapidly northwards, it has raised in the Himalayas the highest mountains in the world, through its collision with Eurasia.

(5) Pacific

A totally oceanic plate, its movement north-westwards, and its collision with Eurasia have produced the trenches and island arcs of the north and west boundaries. Its rate of spreading away from the East Pacific Ridge is some 8 to 10cm a year.

(6) Nazcan

This plate lies between South America and the Pacific plate. Moving eastwards, it is in violent collision with the continent, raising the high mountains of the Andes all the way along the western coast.

(7) South American

The movement westwards from the Mid-South-Atlantic Ridge points to its ancient connection with Africa some 100 million years ago.

(8) Antarctic

This practically stationary plate forms the southern boundary of the five preceding plates.

In addition to these plates there is evidence that Greenland is separating from North America, and that Eastern Asia is separating from Eurasia. It is likely therefore that two further plates exist, the Greenland and the East Asian, making ten global plates in all.

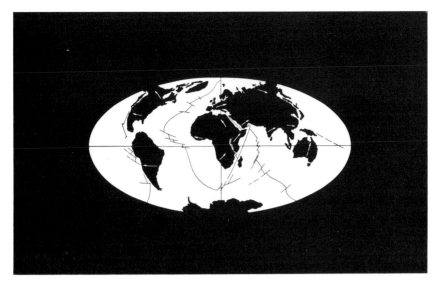

Figure 6.4 – Tectonic plates and their motion   (Geological Museum, London)

6.4 HISTORY OF CONTINENTAL DRIFT

Slowly, a pattern is emerging of continental drift and disruption stretching back at least 700 million years. The studies of paleomagnetism, ancient climates, the evolution and extinction of life forms, the age of mountains, and the evolution of the oceans, yield convincing evidence that the continents have had a rich and mobile history. It will be many years before a detailed account becomes available, but already the broad outlines are clear (Figure 6.5).

In the late Proterozoic, about 700 million years ago, a vast super-continent existed, probably formed from previous separate continents – Wegener's Pangaea. In Cambrian times, about 570 million years ago, Pangaea had split into four continents, roughly corresponding to North America, Europe, Asia and Gondwana. In the seas, covering the continental shelves, invertebrates evolved and diversified rapidly. The number of families increased and then remained constant over a period which stretched from Silurian to Permian times (450 to 250 million years ago). By the Devonian period, 380 million years ago, Europe and North America had collided, raising the Appalachian and Caledonian ranges which ran from Florida, through Wales and Scotland, to the north of Norway. A further unification took place during the Carboniferous period, 300 million years ago, when the sea between Europe and Gondwana closed and the Central Massif, the Harz Mountains, and the Spanish uplands were formed. Seas still separated Asia from the amalgamated continents of Euramerica and Gondwana during the late Carboniferous period, but the gap between Asia and Europe closed, and about 220 million years ago they were

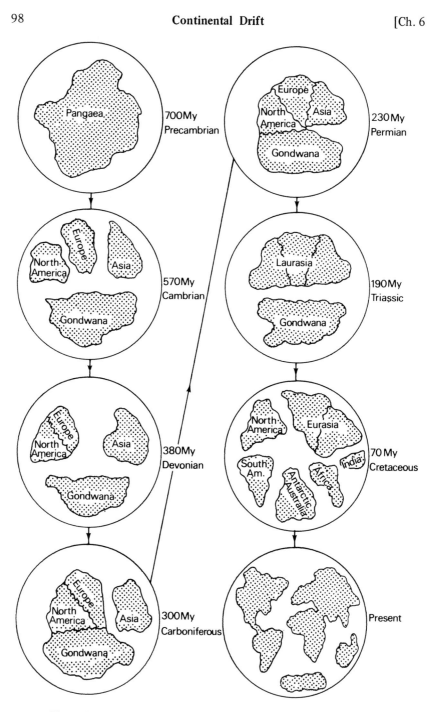

Figure 6.5 — Schematic depiction of history of Continental disruption.

welded together, raising the Urals. This was a catastrophe for sea life, and many species became extinct. The Permian-Triassic boundary marks the end of an era.

The Paleozoic era saw the splitting of Pangaea and its reformation, in a cycle which lasted some 400 million years. The appearance of shallow seas encouraged a great expansion of marine life. As the continents began to move together during Silurian times, sea life came under pressure, and some species evolved the ability to move to the land. Many which did not evolve became extinct when the Pre-Uralian Sea disappeared, and Pangeae was reformed. Yet not all the gaps between the fragments were closed. The Tethys Sea, stretching westwards from what is now east of the Himalayas, across northern Arabia and the Mediterranean, probed deep into the super-continent and separated Asia from Gondwana.

In early Mesozoic times the Tethys Sea expanded, and by about 190 million years ago, towards the end of the Triassic period, the two massive continents of Laurasia and Gondwana had separated: Laurasia consisting of North America, Europe and Asia, to the north; and Gondwana, consisting of South America, Africa, India, Australia and Antarctica, to the south. During the rest of the Mesozoic era, the Jurassic period saw the rise of the dinosaurs, and the beginning of the Cretaceous period heralded the birth of the Atlantic and Indian Oceans. By the end of that period, some 70 million years ago, the dinosaurs were mysteriously extinct, Gondwana had split into South America, Africa, India and Australia-Antarctica, and North America had detached itself from Eurasia. Sea life, however, had once more diversified and proliferated, and mammals had made their appearance.

The recent era, the Kainozoic has witnessed diversification of mammals and flowering plants during the slow drift of continents. Australia separated from Antarctica, India coalesced with Asia, raising the Himalayas. Africa forging north-east, almost closed the Tethys Sea and, continuing the line of mountains westwards from the Himalayas, raised the Caucasus, the Alps and the Pyrennees. The westward drift of the Americas crumpled the Rockies and Andes upwards. The trenches and island arcs of the Pacific came into being. These processes continue to the present day.

6.5 THE GROWTH AND DRIFT OF CONTINENTS

The pattern of continental drift, as outlined above, is known only over the last 700 million years, a comparatively short period of the earth's history. Probably, continental drift has gone on ever since continents existed, but we cannot be sure. Nor can we be sure that the Paleozoic cycle of disruption and unification, in which Pangaea first broke up and then was reformed in a period of some 400 million years, represents a basic oscillatory feature. If it does, then we are today, in mid-cycle, with the continents as dispersed as they will ever get, and in the next 200 million years they will converge, eliminate the shallow seas, and reform Pangaea.

There is, however, a time of this order of magnitude associated with convection in the upper mantle. We can take the distance around the surface of the globe between up-currents, as evidenced by mid-oceanic ridges, and down-currents, as evidenced by subduction, to be about 4,000 km, on average. If convection is confined to the top 500 km of the mantle, material going once round the convection cell makes a trip of some 9,000 km. A drift rate of say 3 cm a year, means that a complete turn-over of material occurs once in about 300 million years, which is of the order of magnitude of the Paleozoic cycle time. This suggests that the inferred period of approximately 400 million years may be intimately associated with convection processes in the upper mantle.

There is another activity which continents indulge in. As well as drifting, they grow. Every continent has a core of very old rocks, dating back to Archaeon times, over 2,600 million years ago. Immediately surrounding the core are younger rocks, some 1,000 to 2,000 million years old, and they are in turn encompassed by rocks younger still, dating from the Cambrian and later periods. This structure, found in the basement rocks, suggests strongly that a continent has grown laterally by the accretian of younger material, which must have come from the mantle. Thus, continents appear to be gradually growing by progressive differentiation from the mantle. Convection in the upper mantle does not merely recycle continental rock, but actively adds to it.

6.6 THE EXPANDING EARTH?

Plate tectonics is not without its critics. One criticism is of particular interest because of its implications for cosmology and for our dependence on the large scale evolution of the whole universe. It has been suggested that plate tectonics does not adequately account for many aspects of geology, in particular sea-floor spreading. The latter points to an expansion of the earth by as much as 10% since Jurassic times, 175 million years ago. The idea that the Earth is expanding goes back to a proposal by Hilgenberg in 1933.

What is so interesting about the idea that the Earth is expanding is that it fits with the predictions of a cosmology based on the variation of fundamental physical constants with time, originally proposed by Dirac in 1938. On the basis of this theory, the gravitational constant G slowly decreases with time, and consequently there should be a slow expansion of the Earth, and indeed of the planets and of the Sun. To see evidence of such an expansion on the Earth's surface is not easy, but on the moon there has been nothing to disturb its surface since the volcanic outpouring of lavas 3000 million years ago. Such vulcanism followed the massive bombardment which produced the Moon's craters, which terminated 3900 million years ago. The Moon shows no large-scale expansion nor indeed any plate movements. The planet Mercury exhibits contraction features but no tensional features which could be associated with expansion. On Earth, indirect evidence has been sought from paleomagnetism, which provides a record of ancient latitude. If the physical separation of two samples of the same geological period remains constant, a comparison of the ancient latitude separation with the present-day one, will show how much the radius of the Earth has changed. No evidence for expansion has yet been obtained by this method.

We have to conclude that the Earth has not expanded significantly, and that certain constraints on the magnitude of the variation of G must therefore be imposed in cosmology. In any case, the crumpling which has raised the Rockies, the Andes or the Himalayas points to contraction as much as sea-floor spreading points to expansion. The results referred to above provide no evidence for contraction either.

FURTHER READING

Hallam, A., 'Alfred Wegener and the Hypothesis of Continental Drift', (p. 88, *Scientific American*, February, 1975).

Heezen, B. C. and MacGregor, I. D., 'The Evolution of the Pacific', (p. 102, *Scientific American*, November, 1973).

Heirtzler, J. R. and Bryan, W. B., 'The Flow of the Mid-Atlantic Rift', (p. 78, *Scientific American*, August, 1975).

Holmes, A., *Principles of Physical Geology*, (2nd edition, Ronald Press, 1965).

Hurley, P. M., 'The Confirmation of Continental Drift' (*Frontiers of Astronomy*, April, 1968).

James, D. S., 'The Evolution of the Andes' (p. 60, *Scientific American*, August 1973).

McElhinny, M. W., Taylor, S. R. and Stevenson, D. J. 'Limits to the Expansion of Earth, Moon, Mars and Mercury, and to changes in the gravitational constant (*Nature*, **271**, 316 (1978).

Stacey, F. D., *Physics of the Earth* (Wiley, 1960).

Toksöz, M. N. 'The Subduction of the Lithosphere', (p. 88, *Scientific American*, November 1975).

Valentine, J. W. and Moores, E. M. 'Plate Tectonics and the History of Life in the Oceans', (p.80, *Scientific American*, April, 1974).

7

Natural Forces

> The motion of each of all the natural species proceeds according to a certain principle. Different species are moved in different ways, and each species always preserves the same course in its motion so that it always proceeds from this place to that place and, in turn, recedes from the latter to the former, in a certain most harmonious manner. We inquire particularly from what source motion receives order of this kind.
>
> Marsilio Ficino (1433–1499)
> *Five Questions Concerning the Mind*

7.1 INTRODUCTION

Our experience of the physical environment is by no means limited to one level, the level at which we feel the warmth of the sun on our faces, or the numbness of frozen fingers, or the buffeting of a gale. There are many other levels of appreciation. We do not feel the earth rotate, or orbit around the sun, but we know that it does. We do not feel continents drifting, but we are aware that they do. Such awareness is part of our common mental environment, the landscape of ideas which represents our understanding of the world. It is never enough merely to experience. We need to extract from the experience an understanding, because this creates order and extends our awareness beyond the raw output of our own sense organs. The most basic conceptual components of our physical environment are the natural forces which move matter, and to which we are all subject.

Physics has identified four basic forces (Figure 7.1). There are two nuclear forces, called the strong and weak interactions respectively. There is the electromagnetic force, and there is gravity. The latter holds us to the earth, and holds the planet to the sun; electromagnetic forces hold atoms and molecules together; and the nuclear forces keep the nucleus of every atom from flying to bits. Matter, whatever its sort, shape or size, also exhibits the property of inertia. Bodies do not instantly change their state of motion when they experience a force. A body likes to travel in a straight line with uniform speed, and resists any attempt to disturb it. Material forced to move around on the surface of a rotating sphere, such as the air and water on the earth, is therefore obliged to exhibit inertial effects continually, and we describe such effects in terms of notional forces like the centrifugal force and Coriolis' force. These inertial 'forces' must be added to the four basic ones.

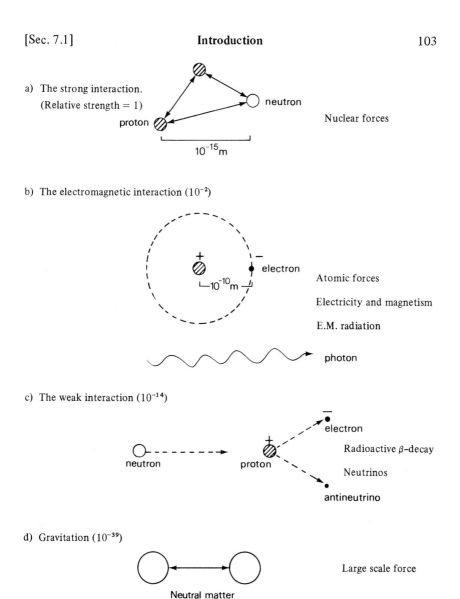

a) The strong interaction.
   (Relative strength = 1)

proton

neutron

Nuclear forces

10^{-15}m

b) The electromagnetic interaction (10^{-2})

electron

Atomic forces

Electricity and magnetism

E.M. radiation

10^{-10}m

photon

c) The weak interaction (10^{-14})

neutron   proton

electron

Radioactive β–decay

Neutrinos

antineutrino

d) Gravitation (10^{-39})

Large scale force

Neutral matter

Figure 7.1 – The four basic interactions.

All of the aforementioned forces act on individual particles. There is, how-ever, another class of effects, whose causes may be regarded as more or less real forces, which appear when large numbers of particles are involved. Examples are the effects of pressure variations and convection. The common feature of such effects is that they are associated with the spatial variation of some quantity, such as temperature, concentration or pressure. Therefore we can conveniently label the causes as 'gradient forces'. The force associated with a pressure gradient is a real mechanical force. On the other hand the force associated with a concentration gradient, producing diffusion, is purely metaphorical, since diffusion comes about because of random motion and requires no force. Bearing in mind that some may be nothing but useful fictions, like the inertial forces, we will add gradient forces to our list. In gravitational torques and tensions, and in the convectional force which drives the continents, we have already come across examples of gradient forces.

7.2 NUCLEAR FORCES

The strongest force we know is the force which binds protons and neutrons together in the nucleus of an atom. Each proton carries an elementary positive charge of magnitude about 1.6×10^{-19} coulomb; and so, being charged alike, protons repel each other electrically. Packing them together in a sphere of radius approximately 10^{-15}m, even though electrically neutral neutrons dilute the packing, entails the existence of an attractive force some hundred times stronger than the electrical one, but having an extremely short range. The binding energy of a nuclear particle is the energy needed to knock a particle out of the nucleus, and it is enormous on the atomic scale. A convenient measure is the energy an electron gains when it is accelerated through a potential difference of 1 volt. This unit is the electron-volt, symbol eV, and it is equal to about 1.6×10^{-19} joules. To remove one of the outermost electrons of an atom takes about 10eV, but to remove a nuclear particle from a nucleus takes about 8MeV (eight million electron-volts). It does not matter much whether the particle is a proton or a neutron. The strong interaction binds both with equal strength.

The nuclear binding energy is not the same for all nuclei (Figure 7.2). It is impossible to overestimate the significance of this fact for the environment. The simplest compound nucleus is that of deuterium, an isotope of hydrogen, the nucleus of which consists of one proton and one neutron. The binding energy per particle is only about 1MeV. Helium has a very stable nucleus, which appears as an α-particle in many radioactivity processes. It consists of two protons and two neutrons, with a binding energy per particle of about 7MeV. The binding energy per particle increases with the number of particles in the nucleus, up to a maximum of 8.79MeV for the nucleus of iron, containing 26 protons and 30 neutrons, and thereafter slowly falls to about 7.6MeV for the nuclei containing the highest number of particles (about 250). The variation pivots about the iron nucleus. If lighter nuclei fuse together or heavier nuclei split up, energy will be released. Nuclear fusion and nuclear fission are the two processes which release some of the vast concentration of energy in the nucleus. How vast this concentration is can be judged from the fact that the release of 1MeV per particle corresponds to some 10^{11} kilojoules per kilogram, whereas the output from burning petrol is only some 5×10^4 kJkg^{-1}, over a million times smaller!

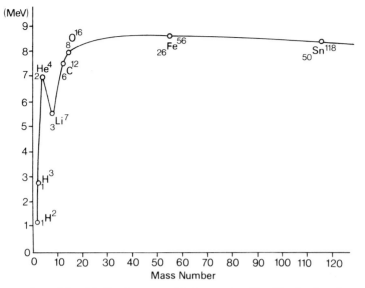

Figure 7.2 – Binding Energy per Nucleon against Mass Number (number of nucleons in nucleus). Explicitly shown are deuterium ($_1H^2$) 1.11 MeV tritium ($_1H^3$) 2.83 MeV, helium ($_2H^4$) 7.07 MeV, lithium ($_3Li^7$) 5.61 Mev, carbon ($_6C^{12}$) 7.68 MeV, oxygen ($_8O^{16}$) 7.97 MeV, iron ($_{26}Fe^{56}$) 8.79 MeV, and tin ($_{50}Sn^{118}$) 8.50 MeV. The curve descends to lead $_{82}Pb^{206}$ 7.88 MeV, thorium $_{90}Th^{232}$ 7.62 MeV and uranium $_{92}U^{238}$ 7.58 MeV.

7.2.1 Energy of the Stars

The fusion of hydrogen to form helium is the source of solar energy on which life on earth depends. Uncontrolled fusion provides the explosive power of the hydrogen bomb, and controlled fusion is the aim of much technological research directed towards the development of a new and powerful source of energy. For protons to fuse they must overcome their mutual repulsion and they will do this only if they collide at sufficiently high speeds. Consequently, fusion of hydrogen requires very high temperatures – about ten million degrees – and this is the order of magnitude reached near the centre of the sun. Moreover the high densities of the solar interior ensure a high frequency of collisions. The difficulty of achieving controlled fusion in the laboratory is in the simultaneous achievement of the high temperatures and high densities, which occur naturally in stellar interiors.

In practice, the heavy isotopes of hydrogen (tritium, which contains one proton and two neutrons in its nucleus, and deuterium, which contains one proton and one neutron in its nucleus) are the working materials. Fusion of a deuterium and a tritium nuclei to form a helium nucleus releases a fast neutron whose energy of about 18MeV can be used to drive turbines and generate electricity. At the time of writing no successful generator has yet been made.

A star can continue to 'burn' its hydrogen for over 10^{10} years. But eventually all interior hydrogen will be used up, leaving an ash of helium. If the star is massive enough it can raise its temperature by contracting internally to trigger off another energy-releasing mechanism — the fusion of helium to carbon and oxygen, which requires a temperature of 2×10^8K. Carbon and oxygen can also be fused at even higher temperatures. Ultimately, the ash of such processes is iron, and no further energy is then available. The prevalence of iron in meteorites and in the earth's core, is therefore no coincidence. It is the end product of fusion processes in massive stars.

Nuclear fusion, therefore, is the source of solar energy and the source of the lighter elements, built out of hydrogen. Elements heavier than iron are thought to be produced in massive stars which go unstable and literally explode. These are the supernovae which ocasionally appear suddenly in the night sky, shine brightly for several months, then die away to become faint, unremarkable, stars.

7.2.2 Radioactivity

The weaker of the two nuclear forces is called the weak interaction, and like the strong interaction is of extremely short range (10^{-15}m). It is the force which is associated with, for example, the radioactive decay of the neutron. A neutron is usually stable within a nucleus, but when free it decays after about quarter of an hour, into a proton, an electron and an antineutrino. In a radioactive nucleus such a decay would produce a fast electron known as a β-particle. Elementary particle interactions which involve that curious particle, the neutrino — a massless thing, with spin, which travels at the speed of light — or its antiparticle (Section 9.8), are all governed by the weak interaction. In strength, it is considerably weaker than the electromagnetic interaction, but much stronger than gravity — the relative strengths of the strong, electromagnetic, weak and gravitational interactions are roughly in the ratios $1:10^{-2}:10^{-14}:10^{-39}$.

Besides being responsible for radioactivity involving β-emission, the weak interaction has a vital role to play in the interior of a star because of its association with neutrinos. Neutrinos interact only weakly with other forms of matter and consequently, when they are produced by a weak interaction, they tend not to be absorbed by surrounding matter. They escape fairly readily even from stellar interiors, and since they carry away energy they constitute an important source of cooling. At temperatures in excess of 10^9K, neutrino production is thought to be so efficient that it constitutes a kind of nuclear refrigeration. This may lead to the collapse of the stellar interior, and this in turn may trigger off a supernova explosion in the star's outer layers, during which many heavy elements are created and blasted into space to form the debris out of which further stars and planets may grow. Through nuclear refrigeration, the weak interaction may well have played a significant part in determining the abundance of the elements found in the earth.

7.2.3 Life is left-handed

The weak interaction may also have played a significant role in determining the structure of life on earth! One of the curious features of the amino acids, which are the building blocks of life, is that they are all left-handed. Their three-dimensional molecular structure has an intrinsic asymmetry, which shows up in their ability to rotate the plane of polarization of polarized light to the left. The odd thing is, that if these materials are synthesised in the laboratory, one gets equal amounts of left-handed and right-handed molecules, a so-called racemic mixture. Life has somehow selected the left-handed variant. Once an initial selection had taken place it is not surprising that this preference would be built-in to the genetic structure, and lead to the complete rejection of the right-handed sort, since the latter would not have the right shape to fit into the complex structure of proteins. Chemically, the two shapes are indistinguishable, so what caused the original selection? An answer to this question has recently been suggested. It is possible that this selection is associated with the lack of symmetry — the so-called non-conservation of parity — found in weak interactions. In radioactive β-decay and in the decay of mesons and other fundamental particles, (all processes involving the weak interaction), it is found that a definite handedness exists in each case. What is striking is the evidence that electrons from such a decay-process, if accelerated towards a target consisting of a racemic mixture, destroys the right-handed component preferentially. If the world consisted of antimatter, the left-handed component, one infers, would be destroyed. Thus it may be conjectured that the left-handedness of the amino acids found in living organisms, stems from the parity violation of natural radioactive processes, and in a hypothetical antimatter world the right-handed form would triumph. The experimental evidence, however, is far from being conclusive at the present time.

7.3 ELECTROMAGNETIC FORCES

Electric forces are responsible for binding electrons to the nucleus in an atom, binding atoms together in molecules, and binding molecules together in solids. Magnetic forces also contribute, but compared with electric forces in most cases they are weak. Electric forces arise out of the repulsion of charges with the same sign, and the attraction of charges with opposite sign. In ordinary matter they are associated, at base, with the positively charged proton and the negatively charged electron, and these particles carry the same magnitude of charge (about 1.6×10^{-19} coulomb). The equality of magnitudes of charge on the electron and proton has been established to within 1 part in 10^{20}, an unusually high degree of precision. Magnetic forces arise out of the movement of charge. Moving charges produce magnetic fields which exert forces on moving charges. They also arise out of the spin of a particle. Because they spin, in some sense, electrons and protons act as tiny magnets. Even a neutron, although it is uncharged, spins and somehow acts as a magnet. The magnetisation of a ferro-magnetic material, like iron, arises because electronic spins line up and reinforce one another's magnetic field.

If charges are accelerated they radiate energy in the form of an electro-magnetic wave (Figure 7.3). Oscillating electrons in a metal aerial radiate radio waves; or, if the frequency of oscillation is high enough (e.g 10^{10}Hz), micro-waves. Our global civilisation produces an appreciable amount of such radiation in its radio, television and radar transmissions. Electrons which make transitions from a high to a low energy state in an atom, or molecule, radiate infrared, visible, ultraviolet or x-ray radiation. Quantum physics has shown that all electromagnetic radiation consists of particles called photons, which have no mass but carry a packet of energy and momentum at the speed of light. Photon energies in the infrared are in the range, roughly speaking, 0.01 to 1.5eV; in the visible they are 1.5 to 3.5eV; in the ultraviolet and x-ray they cover a range from a few electron volts to tens or hundreds of kiloelectron-volts. Beyond these energies are the million-volts of the γ-rays emitted in nuclear processes.

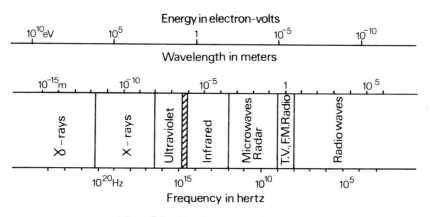

Figure 7.3 – The electromagnetic spectrum.
The shaded region is the visible range.

7.3.1 Thermal radiation

Any chunk of matter, therefore, consists of positively charged nuclei surrounded by clouds of lighter, negatively charged electrons more or less tightly bound to the nuclei, and the whole permeated by photons of all sorts continually being absorbed and emitted. Above the absolute zero of temperature everything jiggles. Jiggling nuclei and photons knock electrons into and out of high energy states, and when electrons get knocked down from high energy states, they emit photons and jiggle the nuclei. At thermodynamic equilibrium a balance is established between the amount of jiggling, the intensity of radiation, and how the intensity is spread over the electromagnetic spectrum (Figure 7.4).

For an ideal material that can absorb and emit all possible photons – we call such a material a black-body – we get a spectrum characteristic only of the absolute temperature, and not at all on the nature of the matter. Although real materials are grey, rather like black-bodies, the black-body radiation spectrum is a very conveneint approximation. Two of its properties are summarised in Stefan's Law and Wien's Law. Stefan's Law states that the total intensity I (energy per second per unit area) is proportional to the fourth power of the absolute temperature T, viz:

$$I = \sigma T^4 , \qquad (7.1)$$

where σ is Stefan's constant (5.670×10^{-8} $Wm^{-2}K^{-4}$). Wien's Law states that the wavelength λ_m of the maximum intensity is inversely proportional to the temperature viz:

$$\lambda_m = C/T , \qquad (7.2)$$

where C is 2.898×10^{-3}mK. From the latter equation we find that around room temperature (290K) λ_m is 10μm, a wavelength in the infrared. We may also note that the peak in the solar spectrum occurs near 0.49μm, which puts the surface temperature of the sun at about 6,000K.

The temperature in these equations is the absolute temperature measured in degrees Kelvin (in tribute to Lord Kelvin). As in the more familiar centigrade scale there are 100 degrees Kelvin (100K) between the freezing point and boiling point of water, but zero degrees 0.0 K corressponds to $-273°$C approximately. The absolute nature of the scale is quite definite. The zero of the centigrade scale is arbitrarily defined as the freezing point of water, and this serves us perfectly well as long as we are interested only in difference of temperature. But absolute zero is not like that. It comes about by taking a gas, keeping its pressure constant, and watching how its volume contracts as we cool it. If we choose a gas like helium which does not liquify until the temperature is very low indeed, we can observe over an enormous temperature range in this way; and we find that the volume is proportional to the temperature. Absolute zero is where the volume would be zero, if the gas did not liquify first. It does not depend on the gas: hydrogen, nitrogen and oxygen all give the same result. Absolute zero is where a gas occupies no space – there is nothing more absolute than that! The absolute scale of temperature is therefore a vital entity, faithfully depicting one of the fundamental ways in which things behave.

Anything above absolute zero radiates electromagnetic energy, and if it has ideal (meaning particularly simple) properties it radiates according to Stefan's Law and Wien's Law: double the absolute temperature and the radiation rate goes up by a factor of 16, and the wavelength of the peak emission halves. Everything around us radiates furiously – mostly in the infrared. Any one thing keeps at that curiously vague level known as room temperature in spite of its intense emission, only by absorbing the radiation produced by others, or acquiring heat by conduction or convection. Ultimately, room temperature is maintained by the absorption of solar radiation, and we will be discussing this in a later chapter. Thermal radiation is a consequence of rotating and vibrating molecules and jiggling atoms and electrons, a consequence of there being, above the absolute zero, thermal energy of motion.

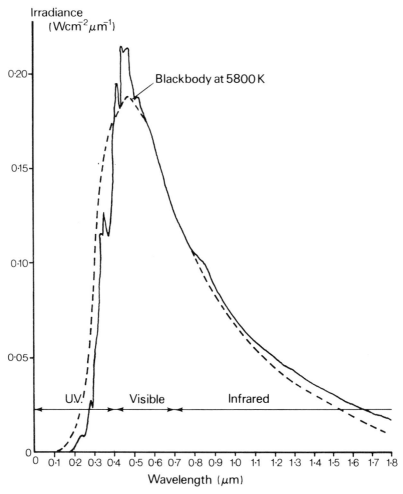

Figure 7.4 – Solar Spectral Irradiance (outside atmosphere) and Blackbody Radiation at 5,800K.

A convenient measure of thermal energy is the energy each vibrational mode of material has, if the total vibrational energy were equally distributed among the many possible modes which can exist. This energy is proportional to the absolute temperature, viz: $E = kT$, where k is Boltzmann's constant. 1.38×10^{-23} JK^{-1}. At room temperature kT is about 0.025eV. As we have seen, the wavelength at the maximum of the blackbody emission at this temperature is $10 \mu m$, and this corresponds to a photon energy of 0.12eV. It is clear therefore that any atomic structures held together with binding energies less than a few tenths of an electron-volt are going to be unstable at ordinary temperatures. Not surprisingly, the strength of the chemical bonds in the materials commonly found in animate and inanimate matter, which as we have said are principally determined by electric attraction, is significantly greater than this, being typically several electron volts. They are therefore stable, but few would remain so if the temperature increased ten-fold. On the other hand, chemical reactions between molecules involve the emission or absorption of energy, which is typically much less than the energy involved in a bond, since the reaction merely exchanges atoms. Thus the energy involved in a chemical reaction is typically a few tenths of an electron-volt. Many chemical reactions can readily take place, therefore, at room temperature, but many reactions vital to living organisms would be inhibited if the temperature dropped by a factor of two.

7.4 THE STRENGTH OF CHEMICAL BONDS

Let us look at this point in a semi-quantitative way, not only in order to explore the situation further, but to illustrate that sort of order-to-magnitude calculation one can do, as it were, on the back of an envelope. A basic result of thermal physics is that the probability $P(E)$ of a thermal process which requires an energy E in order to take place, is proportional to the base of naperian logarithms, e, raised to the power $(-E/kT)$, that is, as the temperature increases, the probability increases exponentially. Heat makes a molecule vibrate vigorously – the higher the temperature, the greater the amplitude of vibration, the more probable the components of the molecule will come apart. Once every vibrational period the bond will be stretched, and so the rate R of dissociation will be proportional, not only to the probability $P(E)$, with E equal to the binding energy, but also to the frequency f of vibration. The more frequently the bond is stretched the greater the rate at which molecules dissociate. A rough value for the frequency of vibration may be obtained by thinking of the oscillation of the molecule as a sound wave of wavelength equal to the bond length. The latter is typically of the order of 1 Å (10^{-10} m) and the velocity of sound in a solid is of the order of 10^3 ms^{-1}. Since in a wave we have the relation between velocity v, frequency f and wavelength λ, given by v $= f\lambda$, f turns out to be of the order of 10^{13} Hz. This is just the order of frequency of infrared radiation which is observed to excite vibrational modes in molecules, so our exceedingly crude calculation is not too far out. Roughly then, a molecular bond is stretched ten million million times a second. If the binding energy E equalled the thermal energy of the vibration, kT, the chances would be that the molecule would dissociate after about three oscillations, corresponding to a dissociation rate of just over 10^{12} s^{-1}, or to a lifetime of just under one picosecond (10^{-12} s). The lifetime reaches one second when E is $30kT$, and roughly one year when $E = 50kT$. A typical binding energy is 3eV. (Table 7.1).

Table 7.1 — Bond Energies

(a) Single Bonds between 'Organic' atoms†		(b) Bond strengths in common materials‡		
Bond	Approx. Energy eV	Material		Energy eV
—O-O—	1.44	Nitrogen	N=N	4.90
>N-N<	1.66	Oxygen	O=O	2.58
>C-N<	3.02	Water	H—OH	5.16
>C-C<	3.60	Carbon dioxide	O=CO	5.50
>C-O—	3.64	Methane	H—CH$_3$	4.51
>C-H	4.28	Ammonia	H—NH$_2$	4.77
H-H	4.51	Formaldehyde	H—CHO	3.77

†Pauling, *The Nature of the Chemical Bond,* 3rd ed. (Cornell University Press, 1960).

‡*Handbook of Chemistry and Physics*, 53rd edition (CRC Press 1972–73).

The rate of chemical reaction between two molecules will depend upon all sorts of factors. Most important will be the frequency of collisions, for it is only when the molecules are close together that a chemical reaction can occur. The collision frequency will depend upon concentration and how fast molecules move around. Molecules in solids may be present in extremely high concentrations but usually cannot move around, so chemical reactions within solids are virtually ruled out. In gases, molecules can move around very easily but are present usually in low concentration, whereas in liquids movement is impeded but the concentration is high. Chemical reactions occur in both gaseous and liquid states.

Not all collisions result in chemical reaction. There is sometimes a barrier to overcome before atoms can be exchanged or added. If this energy barrier is very large compared with the energy involved in the collision, which itself is determined usually by the temperature, reaction will be very rare. Thus the reaction rate can be regarded as the product of two quantities, one being the collision frequency, the other being the probability of reaction, both of which increase with temperature usually. Clearly we cannot say anything general and quantitative about chemical reaction rates since they depend upon so many variables, except to point to the desirability of having liquids and gases, rather than solids, as media in which chemical processes take place.

In the air around us the collision rate is about $10^{10}s^{-1}$. Fortunately, in spite of this high collision rate, the air is highly stable, and this is because nitrogen and oxygen molecules have to overcome an energy barrier of nearly 2eV in order to react, which calls for temperatures of over 2,000K. Such temperatures can occur naturally in lightning flashes. The reaction results in nitric oxide, NO, and other oxides of nitrogen which dissolve in water to give nitric acid. Although useful for restoring nitrates to the soil, to the benefit of the vegetable kingdom, the reaction is somewhat noxious for air-breathers!

7.4.1 Chemical criteria for life

We are now in a position to see how potently the strength of the electric force in atoms and molecules determine the environment. If the environment is to be chemically active, the prevailing temperature must not be so high that few stable molecules can exist, nor so low that chemical reactions are inhibited. Outside of this temperature range chemically-based life (and we know of no other sort) cannot exist. This range is largely determined by the forces of electromagnetism, which determine the strength of chemical bonds.

More subjectively, the range is influenced by our earth-bound ideas of what rates are acceptable. To creatures for whom a microsecond is an age or for whom a thousand years is a mere moment, the problem would look very different. But to us, and our fellow life-forms, with our circadian, monthly and annual rhythms, our common carbon chemistry and our dependence, directly or indirectly, on solar energy, it seems necessary that molecules remain stable for a matter of years, but are to indulge in reactions as rapidly as perhaps a thousand times a second. With these parochial criteria the dissociation rate must not greatly exceed 10^{-7} s^{-1}, and the reaction rate must not fall much below 10^3 s^{-1}. With E = 3eV, the dissociation rate corresponds to an upper temperature of 750K (or about 500°C). If we relax our stability criterion by a factor of a hundred, so that a molecule with a binding energy of 3eV is stable over a day rather than a year, the upper limit rises to about 1,000K. But we cannot push the lower limit down much below say, 250K, on the grounds that very few polyatomic molecules remain liquid, and chemical reactions between solids occur only very slowly. A temperature of 200K, which is just above the points at which carbon dioxide (CO_2) and ammonia (NH_3), two inorganic compounds closely associated with biological activity, become solids, is about the lowest limit.

An objective criterion of the upper limit of the temperature range may be obtained by identifying this limit with the temperature at which the dissociation rate equals the reaction rate. At this temperature molecules dissociate as rapidly as they react, and chemistry disappears. Chemically-based life, however, unfamiliar, cannot exist under such conditions. Again taking the binding energy E to be 3eV, and equating the dissociation rate with the collision rate, we get a temperature of some 6,000K.

If we tie the possibility of life with the possibility of chemical processes we see that the environment must offer a temperature between 200 and 6,000K. Since the latter is of the order of the surface temperature of our sun and many other stars, it is probable that life can only exist on planets, and only on those warmed to at least 200K by their local star. The range of temperature seems large, but it is narrow compared with the range from near 0K in space to over 10^9K in the interior of massive stars. This cosmically narrow range, so important for life, is a remarkable consequence of electromagnetic interaction.

7.4.2 Water: the basis of life?

In taking the criterion for the existence of life to be that for the existence of chemistry, we have deliberately adopted an extremely non-parochial viewpoint. Even so, this approach has led to a very narrow range of temperature in which chemical life — taking a broad sense of life — may possibly occur. It may turn out that a parochial view is really more accurate in the universe at large. It may be that carbon chemistry, and no other, is the only possible basis for life. Moreover it may be that the existence of the simplest liquid containing the most abundant element hydrogen — namely, water — is absolutely essential. We know too little about life-forms in the universe at large to be sure about such points, and we are very aware of the dangers of unimaginative extrapolation from what we know about on earth. Nevertheless, if it is indeed the case that water is a requirement, the temperature range narrows abruptly from 200K to 6,000K right down to 273K–373K at standard atmospheric pressure. As it happens, taking water to be a requirement is not all that parochial. Table 7.2 shows the melting points and boiling points at standard pressure of hydrogen compounds containing the common, lighter elements, and it is clear that water is remarkable.

Table 7.2 – Melting and Boiling Points of Hydrogen Compounds containing common, light elements†

Substance	Formula	M.P.(K)	B.P.(K)
Methane	CH_4	91	109
Ammonia	NH_3	195	240
Hydrogen Cyanide	HCN	259	299
Water	H_2O	273	373
Hydrogen Fluoride	HF	190	293
Silane	SiH_4	88	161
Phosphine	PH_3	140	185
Hydrogen Sulphide	H_2S	187	212
Hydrogen Chloride	HCl	158	188

†Handbook of Chemistry and Physics 53rd edition 1972–73 CRC Press.

7.5 GRAVITATION

Gravitation, the force of attraction which matter exerts on matter, is by far the weakest of all interactions. Compare the electric attractive force between a proton and an electron, which holds the hydrogen atom together, with the gravitational force between these particles. It is an exceedingly large number, of order 10^{40}: the intrinsic strength of gravitation is forty powers of ten smaller than the electric interaction. Nevertheless, it is the dominant force at the cosmic level, basically because positive and negative charges in matter neutralise one another so effectively.

We have already discussed the role of gravitational torques and tensions in slowing the earth's rotation (Section 3.5), and we have mentioned how gravity and inertial forces effect the earth's shape (Section 4.1). Gravity's central role in celestial mechanics needs no emphasis — the earth's orbit around the sun, the perturbations of that orbit, make the varying pattern of gravitational fields in the solar system an important part of our physical environment. Here, we will limit ourselves to mentioning just one aspect of the earth's gravitational field, an aspect to do with imprisonment.

Because of the gravitational attraction it is difficult for matter to escape from the earth. It can free itself only by acquiring a high enough velocity. The lowest velocity which frees a body from the surface is one which is associated with a tight orbit around the earth, and is directed horizontally (see Section 4.1). In order to skim around the earth without falling one must travel at a velocity of 7.9kms^{-1}, or once around the earth in 84 minutes (Figure 7.5).

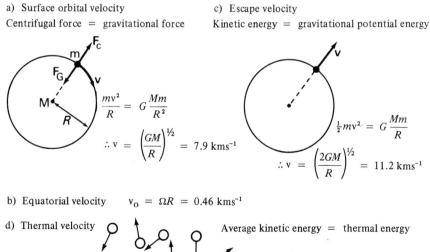

a) Surface orbital velocity
Centrifugal force = gravitational force

$$\frac{mv^2}{R} = G\frac{Mm}{R^2}$$

$$\therefore v = \left(\frac{GM}{R}\right)^{\frac{1}{2}} = 7.9 \text{ kms}^{-1}$$

c) Escape velocity
Kinetic energy = gravitational potential energy

$$\tfrac{1}{2}mv^2 = G\frac{Mm}{R}$$

$$\therefore v = \left(\frac{2GM}{R}\right)^{\frac{1}{2}} = 11.2 \text{ kms}^{-1}$$

b) Equatorial velocity $v_0 = \Omega R = 0.46 \text{ kms}^{-1}$

d) Thermal velocity

Average kinetic energy = thermal energy

$$\tfrac{1}{2}mv_T^2 = \tfrac{3}{2}kT$$

$$\therefore v_T = \left(\frac{3kT}{m}\right)^{\frac{1}{2}}$$

Figure 7.5 — Velocities

To escape totally, a body must acquire a vertical velocity such that its kinetic energy is at least equal to the gravitational potential energy at the surface. The body will gradually slow down as it moves away, but it must reach zero velocity only at infinity. Directed vertically the velocity must be 11.2km s^{-1}, which is exactly $\sqrt{2}$ times the skimming velocity.

To escape requires a velocity of order 10kms^{-1}, or some 40,000kmhr^{-1}. Compare that with a walking speed of 1ms^{-1} (a factor of 10^4 too slow), a car travelling at 100km hr^{-1} (a factor of 400 too slow), and a supersonic jet travelling at 4,000km hr^{-1} (a factor of 10 too slow). It has needed gigantic rocket motors to lift man off his planet, and even so there has been so little momentum left over, that it has taken him several days to cover the 400,000km to the moon.

But there are naturally occurring fast objects all around us. The thermal energy of a gas manifests itself in the random motion of the molecules composing the gas. Naturally, a wide range of velocities occur. The kinetic theory of gases describes the distribution function, that is, the probability of a molecule having a given velocity, and it also gives us a simple relation between the average velocity of a molecule and the absolute temperature of the gas. As long as quantum effects are negligibly small, the average thermal energy in any mode of motion is just $\frac{1}{2}kT$, where k is Boltzmann's constant, and T is the absolute temperature. In straightforward translation motion there are three possible modes, each corresponding to one space dimension, and so this sort of motion has, on average, a kinetic energy of $\frac{3}{2}kT$. The average thermal velocity corresponding to this energy is readily calculated (Figure 7.5). At 250K, the average stratospheric temperature ($-23°$C), a hydrogen molecule (the lightest, and therefore the fastest) has an average thermal velocity of 1.8 kms^{-1}. Nitrogen and oxygen molecules move some five or six times more slowly. On average, no molecule in the atmosphere has the velocity to escape, not even hydrogen. The atmosphere is safely held prisoner. However, the average velocity of hydrogen molecules is only a factor of about 7 smaller than the escape velocity. A small but significant fraction of molecules will have velocities high enough to allow the gas to escape over a period of time. Thus hydrogen gas cannot be a stable component of the earth's atmosphere — it escapes too easily. Similar remarks apply to the next lightest gas, helium. After helium the lightest gases are nitrogen and oxygen, the molecules of which are heavy enough to escape only slowly. Above 100km, however, gases become stratified according to their molecular weight, and so at a height of 1,200km hydrogen becomes in fact the dominant constituent of the atmosphere which by then is extremely rarified. The lower atmosphere is dominated by nitrogen and oxygen. These gases are the major components of the atmosphere. Hydrogen and helium are the most abundant elements in the universe, but only planets substantially colder or heavier than the earth can retain them in free gaseous form. The planets Jupiter, Saturn, Uranus and Neptune are all able to retain substantial quantities of hydrogen, so much so that hydrogen is perhaps the dominating component of the atmospheres of Jupiter and Saturn. But on earth, hydrogen is retained substantially through chemical combination with oxygen as water, and very little is free in the atmosphere.

7.6 INERTIAL REACTIONS

The response of matter to a force is determined by something we call inertia. The natural state of motion of a material thing is to travel along in a straight line with uniform speed. Left alone, a body will continue in uniform motion indefinitely. Any deviation from uniform motion is attributed to the presence of a force. Inertia determines how big a deviation from uniform motion in a straight line a force of a given strength will produce. The simplest measure of non-uniform motion is acceleration, the rate of change of velocity; and the amount of acceleration (a) which a body undergoes is determined by the strength of the force (F) and something we call the inertial mass (m) of the body, according to Newton's Law, $F = ma$. The inertial reaction of a body to a force, whether the force be gravitational, electromagnetic or nuclear, is therefore vital for determining motion, and it is a vital component of the physical environment.

The above is nothing more than a brief summary of Newtonian mechanics. Fortunately we often find, in the world about us, that both quantum, and Einstein's relativistic, modifications of Newtonian theory are negligibly small; and so we can, without incurring serious error, employ classical mechanics. Even so, it has to be appreciated that motion is a relative thing – (how relative, is a question still discussed today) – and we may wish to describe motion with reference to a coordinate system which is not the 'fixed stars' system of simple problems. After all, we ourselves are rotating with the earth. It is all very well to describe the track of a bullet along the earth's surface from the stand-point of the Andromeda nebula, but it is more useful to describe it from the stand-point of the shooter or his target, both of which are in the rotating frame of the globe. When we describe inertial reactions to real forces from the standpoint of a rotating frame of reference we find that we have to introduce accelerations, which would not be there if we adopted the so-called inertial frame of reference based on the fixed stars. To explain these accelerations we introduce the conceptions of two forces. One is the centrifugal force, which apparently pushes a body away from the rotation axis; the other is the Coriolis force (named after its Italian discoverer), which apparently pushes at right-angles to the direction of motion of a body and causes its path to be curved. These forces are not like the gravitational or electromagnetic forces. Gravitational and electromagnetic forces do not completely disappear if we shift our description from one based on accelerating frame, to one based on a non-accelerating (inertial) frame. Centrifugal and Coriolis forces do. Nevertheless they describe very real effects.

7.6.1 Inventing inertial forces

Actually, we do not even have to go so far as a rotating frame in order to see how these forces appear. It is enough to consider the unimpeded motion of a body in a straight line from the standpoint of a stationary observer! Even though the body is travelling in a straight line with constant velocity, a stationary observer can easily invent accelerations for it. Let the observer be at O (see Figure 7.6) and let the body travel in a straight line with uniform velocity v. Its closest approach to O is the point A. Let the distance OA be r_0. Some time after passing A the body is at B, a distance r away from O. Obviously r is bigger than

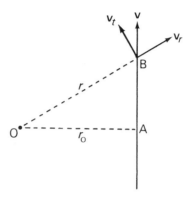

a) Resolve velocity into radial and transverse components

$$v^2 = v_t^2 + v_r^2$$
$$\therefore v_t = v[1 - (v_r/v)^2]^{1/2}$$

Suppose B close to A such that $v_r \ll v$, then

$$v_t \approx v[1 - \tfrac{1}{2}(v_r/v)^2]$$

b) Relation between tangential and radical accelerations is given by differentiating with respect to time keeping v constant, viz.;

$$\frac{dv_t}{dt} = - \frac{v_r}{v} \frac{dv_r}{dt} \qquad\qquad 1$$

c) Angular momentum about O remains constant:

$$mv_t r = mvr_0$$
$$\therefore v_t = \frac{r_0}{r} v$$

$$\therefore \frac{dv_t}{dt} = - \frac{vr_0}{r^2} \frac{dr}{dt} = - \frac{v}{r} v_r \qquad\qquad 2$$

∴ From 1 and 2
$$\frac{dv_r}{dt} = - \frac{v^2}{r} \qquad \text{(Coroilis acceleration)}$$
(Centrifugal acceleration)

Figure 7.6 – Inertial accelerations. Zero rotation.

r_0; the body has moved further away from O. But our observer would have observed that, when the body was at A, its velocity had no component directed away from O. And yet at B the body is further away. Clearly, a component of velocity directed outwards from O has grown during the passage of the body from A to B. The appearance of this radial velocity accounts for the increased separation. But the growth of this radial velocity means that there is a radial acceleration; an acceleration, that is, directed outwards from O.

There we have our first invented acceleration. Because the total velocity is a constant, the growth of its radial component must be compensated by a diminution in the component directed at right-angles to the radial line OB. The tangential velocity, in other words decreases. Here is a second invented acceleration, this time directed at right-angles to the radial motion, continually reducing the tangential velocity, acting in a clockwise sense about O. Thus by merely describing the motion from the standpoint of a stationary observer situated at the fixed position O we invent radial and tangential accelerations. Since accelerations imply forces, we have also invented a radial force – which is the familiar centrifugal force – and a tangential force – which is the Coriolis force in everything but magnitude. To O, the uniform motion of a particle in straight line is the result of an interaction of these two forces.

To describe the uniform motion of a body from A to B in this way seems ridiculously complicated, to say nothing of being egocentric. To the maniac at O, things either stay still or move; and if they move it matters only whether they move towards him or away, or stay at the same distance rotating around him. At A the body seemed to be moving around keeping a fixed distance r_0 from him, in which case, as far as he is concerned, there are no forces acting. Instead, the body moves away and at the same time rotates about him more slowly. Obviously a centrifugal force has pushed it away and a tangential brake has been applied. To us there is a simpler description which involves no forces at all, but to our self-centred observer his description seems more useful, and involves less abstraction.

7.6.2 The rotating-disc world

He becomes much less of a maniac if the world is rotating at a constant rate. Imagine him sitting at the centre of a rotating disc (Figure 7.7). The only change is that bodies which are stationary with respect to him are, to us, rotating with the disc at constant rate. If our body stayed at A our observer would immediately interpret that as a rotation about O in the opposite sense from which we see the disc rotating. Its velocity around him would depend upon the distance away from O, the bigger r, the bigger the velocity. A body at B stationary to us would appear to O to rotate with the same angular speed as our stationary body at A. Since B is further away from O than A, the body at B would therefore have to cover more ground in one revolution than A. Its tangential velocity, in other

words, would be higher. So even if our body travels infinitesimally slowly from
A to B, from our point of view, the rotating observer will see it pick up tangential
velocity and attribute tangential acceleration. This has to be added to the tangential acceleration he observes when the disc is stationary. For relatively slow
motions it turns out that the effect of rotation is just to double the tangential
acceleration, and leave the radial acceleration exactly the same. Things moving
relatively to objects on a rotating disc therefore appear to experience a tangential
force, which we call the Coriolis force, and a radial force which we call the
centrifugal force, both of which depend upon the relative velocity.

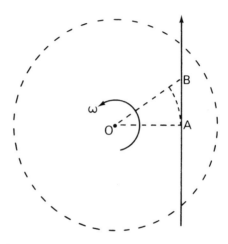

A body at rest relative to the rotating disc has tangential velocity ωr. If the body
moves away from the centre it will remain at rest tangentially relative to the
rotating disc if its tangential velocity increases with an acceleration of $\omega \dfrac{dr}{dt} = \omega v_r$.

To obtain acceleration relative to rotating disc subtract ωv_r from tangential
acceleration relative to stationary disc:

$$\frac{dv_t{}'}{dt} = -\frac{v}{r} v_r - \omega v_r$$

For small velocities relative to rotating disc

$$\frac{v}{r} \approx \omega$$

$$\therefore \frac{dv_t{}'}{dt} \approx -2\omega v_r \text{ Coriolis acceleration on rotating disc.}$$

Figure 7.7 – Coriolis acceleration on rotating disc.

7.7 THE EARTH'S CORIOLIS FORCE

Tackling the problem of motion on a rotating sphere like the earth, is complicated by having three dimensions to worry about and an all-pervading gravita-

tional force (Figure 7.8). Fortunately the two new aspects neutralise one another when we restrict our attention to motion about the surface, and we return to the disc-like situation in which we have a tangential (that is east-west) acceleration and a radial (that is north-south) acceleration, both of which depend upon the latitude. If we combine vectorially these two components of surface acceleration, we obtain a single acceleration directed exactly at right-angles to the direction of the relative motion, and we speak of this as being caused by the earth's Coriolis force. Given that the motion is restricted to the surface, all the inertial effects manifest themselves in this curious tendency of moving things to accelerate perpendicularly to their motion. It is no use telling us that a simpler approach is to look at the motion of things from the stand point of the fixed stars. We live on earth and orientate ourselves by observing objects stationary with respect to the earth. It is usually simpler in fact to work with Coriolis force rather than take an abstract astral viewpoint.

The magnitude of Coriolis force is given by the formula, $F_c = 2mv \, \Omega \sin \phi$ (Figure 7.8). In this expression m is the mass of the body, v is its relative velocity, Ω is the angular frequency of the earth's rotation and ϕ is the latitude. Thus, at the equator $\phi = 0$, and since $\sin \phi = 0$, Coriolis force vanishes. Coriolis force has maximum strength at the poles where $\phi = 90°$ and $\sin \phi = 1$.

We may not be conscious of Coriolis force when we walk or ride about, but it is there all the time. Indeed, it is of such vital importance in under-standing the winds of the atmosphere and the currents of the oceans that we had better be sure we understand how it works. It is all very well tracing its develop-ment in terms of maniac observers and rotating discs, but that sort of thing seems a long way removed from things on earth.

The simplest way of seeing how the earth's Coriolis force comes about is perhaps to recall that angular momentum — the product of momentum (mv) and distance from the rotation axis — is conserved in the absence of torques, and that when a body is at rest on the earth's surface the gravitational force, the centrifugal force, and the reaction of the material on which the body rests, are all in equilibrium. If a body travels towards a pole, it moves from a latitude where the angular momentum is high to one where it is low. Carrying this high angular momentum with it means that it will rotate about the earth's axis faster than it should, and so this will manifest itself in a displacement to the east. Conversely, a body moving towards the equator cannot keep up with the more rapid speed associated with the rotation and lags westwards. Both cases result in a displacement to the right of the motion in the northern hemisphere (to the left in the southern hemisphere). On the other hand, no change in angular momentum is involved when motion is east-west. In this case the Coriolis effect arises out of the change in centrifugal force. Motion eastwards means that the body has a higher rotation speed than the earth and therefore suffers an increased centrifugal force, with the result that it experiences a displacement towards the equator. Conversely, motion westwards reduces the centrifugal force, and gravitation pushes the body towards the pole. Again the Coriolis displacement is to the right in the northern hemisphere (to the left in the southern hemisphere). In general, both angular momentum and centrifugal effects operate. The net result is a moving body always experiences a force at right angles to its motion, whatever direction that motion takes. We reach the conclusion that the natural path followed by a moving object on the earth's surface is a circle.

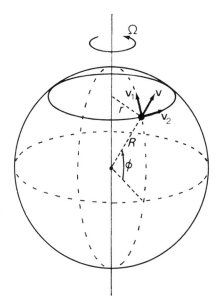

a) Latitudinal component v_2 gives rise to increase in centrifugal force

$$F_c = \frac{mV^2}{r}, \ (V = \omega r),$$

$$\frac{dF_c}{dV} = \frac{2mV}{r} = 2m\Omega$$

$$\therefore \text{Extra force} = 2m\Omega v_2$$

$$\text{Horizontal component} = 2m\Omega v_2 \sin\phi = F_1$$

b) Longitudinal component v_1 gives radial component $-v_1 \sin\phi$ and hence 'disc' Coriolis' force $= 2m\Omega v_1 \sin\phi = F_2$.

c) Total Coriolis' force $=$ vector sum of above components

$$= (F_1^2 + F_2^2)^{1/2}$$

$$= 2m\Omega v \sin\phi$$

Directed at right angles to v.

Figure 7.8(a) – Earth's Coriolis force.

Figure 7.8(b) (*following page*) – Foucault's pendulum at the Pantheon Paris
(Photo: Science Museum, London)

7.7.1 Inertial Oscillations

This is a rather remarkable feature of our environment. Because of the inertia of matter and because of the rotation of the earth, unimpeded motion on the earth's surface is circular. Such a motion is called an inertial oscillation (Figure 7.9). It is a simple fact that the frequency of such an oscillation depends only on the angular frequency of the earth and the latitude as shown by the following formula, $\omega_c = 2\Omega \sin \phi$.

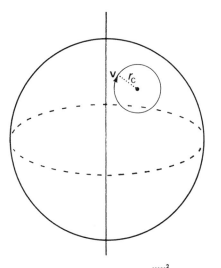

Centrifugal force associates with inertial motion $= \dfrac{mv^2}{r_c} = mv\omega_c$ where angular frequency of oscillation $= \omega_c$.

$$mv\omega_c = 2m\Omega v \sin \phi$$
$$\therefore \omega_c = 2\Omega \sin \phi$$

Figure 7.9 — Inertial Oscillation.

The velocity does not enter into it. To be sure, the larger the velocity, the bigger the radius of the circle, but then the faster is the circumference traversed, and so the time taken to go once round (whose reciprocal is the frequency) does not depend on speed. Thus one oscillation takes 12 hours at the poles, 24 hours at latitude of 30°, and increases without limit towards the equator.

The simplest way of seeing Coriolis' force in action is to watch the motion of a very long pendulum, like the one originally used by Foucault in Paris. With each swing the pendulum experiences a force directed perpendicular to and to the right (in the northern hemisphere) of its motion. The plane of oscillation slowly rotates clockwise. (Figure 7.8(b)).

Of course, motion is usually impeded by friction or viscosity or just plain obstruction, so it is rare to observe this circular motion in its pure form. Nevertheless, though modified by various forces and associated with other vibrations, inertial oscillations are important features of the complex motion of the fluids of the oceans and of the atmosphere.

7.8 GRADIENT FORCES

The previous sections have been concerned with forces experienced essentially by a single particle. When there is a large number of mutually-interacting particles, such as occur in a gas or fluid, a new category of forces enters, which contains interactions of a statistical nature. The motion of molecules in a fluid causes them to collide with one another, and with the material which determines the boundaries containing the fluid. These collisions are, at base, electromagnetic interactions, on a sub-microscopic scale, and when there are a large number every second, the average effect is describable in terms of pressure – the force per unit area which one bit of fluid exerts on an adjacent bit, or on the boundaries.

The average kinetic energy of the molecular motion itself is measured by the absolute temperature T, of the fluid. When quantum effects are negligible, the average kinetic energy of a freely-moving molecule is just $\frac{3}{2}kT$, where k is Boltzmann's constant. Temperature and pressure are two of the three principle parameters which characterise a large collection of interacting particles, and, being statistical quantities, they have meaning only when applied to large assemblies. The third parameter is the number of particles in unit volume – the number density, or its closely associated parameter – the mass density.

In a perfectly uniform fluid none of these quantities in itself gives rise to net forces or movement. Where the pressure is uniform, as much force is extended on an element of the fluid in one direction as in the opposite direction. Where the temperature is uniform, no forces are exerted by differential thermal expansion. Where the number density is uniform (assuming a single chemical type of molecule), as many molecules diffuse to the left as they do to the right. Only where uniformity ends, as at a boundary or, in general, where a variation in space exists, do net forces or motion appear.

7.8.1 Pressure forces

One of the most important of these forces in our environment is that associated with a pressure gradient. Imagine a small cubic volume of fluid (Figure 7.10) where there is a pressure gradient in the x direction. Force is pressure multiplied by area and so the net force will be the difference in pressures multiplied by the area of the side. If $P(x_2)$ is greater than $P(x_1)$, then with the usual convention that x_2 is greater than x_1, we speak of a positive pressure gradient. In such a case the net force is to the left, i.e. in the negative direction, and therefore the net force is proportional to the negative of the pressure gradient. This illustrates a general feature of non-shearing gradient forces, namely, that they point in the opposite direction to the gradients themselves. Mathematically, the pressure gradient is written dP/dx, and the net force per unit volume of fluid is written $\mathcal{F} = -dP/dx$.

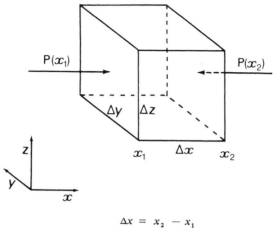

$$\Delta x = x_2 - x_1$$

$$P(x_2) = P(x_1) + \frac{dP}{dx} \Delta x$$

$$\text{Force on left} = P(x_1) \Delta y \Delta z$$

$$\text{Force on right} = -(P(x_1) + \frac{dP}{dx} \Delta x) \Delta y \Delta z$$

$$\text{Net force} = -\frac{dP}{dx} \Delta x \Delta y \Delta z$$

$$\text{Net force per unit volume } \mathcal{F}_p = -\frac{dP}{dx}$$

Figure 7.10 – Pressure Gradient Force

7.8.2 Temperature forces

Temperature gradients give rise to thermal currents, heat flowing from high to low temperatures. But they are also responsible for mechanical forces associated with thermal expansion of concentration. In solids such forces can lead to disruption – part of the cause of rock erosion, and the wear and tear on motorways, especially if icing is involved. In fluids, these forces acting in a gravitational field, are responsible for convection currents – hot fluid expanding becoming less dense, and rising. As such they are a vital factor in determining the circulation of the atmosphere, and indeed the climate itself. The practical importance of temperature gradients therefore, is that they tend to produce density gradients and hence convection, but the relationship between the two quantities, and what motions result, is usually a complicated problem to solve, and dependent on the physical properties of the material and its surroundings.

7.8.3 Random motion

Finally, let us mention the phenomenon of diffusion. Where there is a spatial variation in the concentration of a given chemical or physical species, a diffusion current flows to equalise the concentration everywhere. This current arises not through the action of a force, but simply by virtue of the random thermal motion of the particles. At any point in space where a concentration gradient exists the number of particles per second travelling randomly in one direction will not in general equal the number of particles per second travelling randomly in the opposite direction. More particles will flow from where their concentration is higher, and so a net current of particles will flow in the direction of low concentration. A simple relationship between the diffusion current and the concentration gradient often holds. If j_D is the diffusion current-density (number of particles flowing through unit area per second) and n is the concentration (number of particles per unit volume) then, in one dimension $j_D = -D dn/dx$, where dn/dx is the concentration gradient, and D is a quantity known as the diffusion coefficient. This equation is often referred to as Fick's Law. The diffusion coefficient is dependent on how. rapidly the particles move, and is therefore a function of temperature, the frequency of collisions, and the material through which the particles move. Atoms will diffuse through a gas or a liquid more rapidly than through a solid, and in general more rapidly at high temperatures than at low temperatures. The negative sign denotes that the diffusion current is in the direction opposite to that of increasing concentration.

7.9 WINDS

Pertinent examples of the interplay of inertial and gradient forces are the atmospheric winds and the oceanic currents, both vital features in the workings of the vast planetary heat-engine which determines climate. Driven by the sun's radiation, the engine works in a never-ending task of trying to equalise temperatures over the whole surface, using the atmosphere as its principle working fluid. Heated air rises, cools, and water vapour condenses into clouds; cold air subsides, warms, and being dry, produces fine cloudless conditions. Cold polar air moving south, and warm tropical air moving north, produce the 200km hr^{-1} high-altitude jet streams and their associated complex of turbulence of alternate high and low pressure systems, which dominate the climate of the temperate latitudes. Oceanic currents, like the Gulf Stream, transfer heat from one part of the globe to another, and add to the intricacy of the weather machine.

The task of unravelling the interconnections which exist in the weather-making process is a formidable one. Meteorology may tell us that fine weather is associated with rising atmospheric pressure and descending air, and that cloudy, wet weather is associated with falling pressure where air is ascending, but what gives rise to such conditions is altogether a more difficult problem. For our immediate purpose it will be sufficient to note that regions of high or low pressure occur in the atmosphere. Where there is a pressure gradient, air will be forced into motion, producing a wind.

On a non-rotating earth, a local region of high pressure would produce winds directed down the pressure gradient, away from the centre. Because the earth rotates, these winds are bent by the Coriolis force to the right in the northern hemisphere (Figure 7.11). In the absence of friction — a condition approached more closely at high altitudes than at ground level — the winds blow, on average, at right-angles to the pressure gradient, that is, parallel to the isobars. The winds tend to follow a path *around* a local high-pressure area rather than *away* from it. This circulation, viewed on a weather-map in the northern hemisphere, is clockwise. Such a local high-pressure area is known as an anticyclone, since the circulation is in the opposite sense to the earth's rotation which is termed cyclonic. A similar circulation of air surrounds a local low-pressure area, but it is in the opposite direction, namely anti-clockwise on the weather map. Such a condition is known as a cyclone. In either case the wind, unhampered by obstacles on the ground, blows parallel to the lines of equal pressure. To identify the direction of the pressure gradient, stand with your back to the wind and the high pressure will be to your right (in the northern hemisphere).

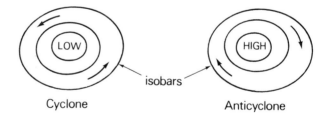

Figure 7.11(a) — Cyclonic and anticyclonic circulation (Northern hemisphere).

Let us look at this in a more quantitative way. Imagine a cubic metre of air in an anticyclone, situated in the northern hemisphere, and suppose that it is initially stationary. It immediately experiences the pressure-gradient force but, being stationary, it experiences no Coriolis force. As a result, it is accelerated down the pressure gradient (Figure 7.12). As soon as it acquires velocity, the Coriolis force acts upon it and induces an acceleration at right-angles to its motion. The packet of air bends to the right, and continues to bend more as it accelerates. Eventually it moves at right-angles to the pressure gradient, but by this time it is moving so fast that the Coriolis' force exceeds the pressure-

Figure 7.11(b) (*above*) – Cyclone
with cold front stretching across
England on 16 Sept. 1978.
(Photo in infrared by satellite
NOAA/5, ESSA, processed by
University of Dundee)
Figure 7.11(c) (*right*) – Cyclone
off Morocco. The photograph also
illustrates geological structure.
                              (NASA)

gradient force and so it is driven back up the pressure slope. It therefore deceler-
ates and the curvature of its path gradually lessens as the Coriolis' force weakens.
Eventually the pressure gradient brings it to a standstill at some position to the
right of where it started, and the whole looping motion begins again. On average,
therefore, the packet of air moves parallel to the isobars.

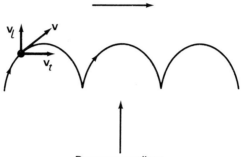

Equations of motion of pocket of air, density ρ, Coriolis frequency ω_c.

$$\rho \frac{dv_\varrho}{dt} = \mathcal{F}_p - \rho\omega_c v_t \qquad\qquad 1$$

$$\rho \frac{dv_t}{dt} = \rho\omega_c v_\varrho \qquad\qquad 2$$

From 1
$$v_t = \frac{\mathcal{F}p}{\rho\omega_c} - \frac{1}{\omega_c}\frac{dv_\varrho}{dt}$$

$$\therefore \frac{dv_t}{dt} = -\frac{1}{\omega_c}\frac{d^2v_\varrho}{dt^2}$$

Substitute in 2
$$\frac{d^2v_\varrho}{dt^2} = -\omega_c{}^2 v_\varrho$$

On average $\dfrac{dv_\varrho}{dt} = 0$.

$$\therefore v_t = \frac{\mathcal{F}p}{\rho\omega_c} \quad \text{on average.}$$

Figure 7.12 — Geostrophic Wind

The motion is not a free, unimpeded motion since there is a pressure gradient
acting. That is why the packet of air performs this looping motion rather than
follow a circular path. Nevertheless, the frequency of looping is just the local
frequency of inertial oscillation, and the trend of the motion is at right-angles
to the pressure gradient with a velocity, averaged over a loop, sufficient to give
a Coriolis' force which exactly opposes that the pressure gradient. Thus, on
average, a flow of air parallel to lines of equal pressure (isobars) is produced.

Such a flow of air, whose average velocity is just the right value in order that the Coriolis' force balances the pressure-gradient force, is known as the geostrophic (literally, earth-turning) wind. In reality, friction with the ground causes the wind to blow somewhat down the pressure slope, and to have a smaller speed, then predicted by the geostrophic windspeed formula $v = \mathcal{F}/(\rho\omega_c)$, where $\mathcal{F}$ is the pressure-gradient force per unit volume, ρ is the density of air, and ω_c is the angular frequency of the inertial oscillation.

7.10 THERMAL WINDS

When cold and warm air lie along side each other, as they do in a frontal system and in general as they do as polar and sub-tropical air masses, another type of wind has to be added vectorially to the geostrophic wind. This wind component is termed the thermal wind, and is associated with a horizontal temperature gradient. In Section 8.3 we will see that the magnitude of the *vertical* pressure gradient decreases as the temperature increases: pressure falls off with height more slowly in warm air than in cold air (Figure 7.13). Thus

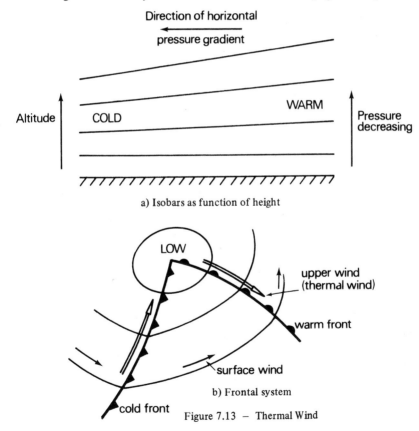

a) Isobars as function of height

b) Frontal system

Figure 7.13 – Thermal Wind

even if no pressure difference exists at ground-level, a horizontal pressure gradient appears and grows with height, the warm air acting as a high pressure zone, and this generates a geostrophic component of wind which blows parallel to the lines of equal temperature (isotherms) and increases in strength with height, as far as the tropopause 10km to 20km high. This is the thermal wind.

The thermal wind associated with the polar-sub-tropical temperature gradient is always on average westerly and manifests itself as the circumpolar jet stream which snakes around the pole in a wavy manner. In frontal systems the thermal wind consists of a high-altitude jet running along the line of the front. Sometimes these upper winds are made visible by cirrus clouds and their direction can be seen by the cloud motion. We can tell something about the front. Stand with your back to the upper wind (which is usually a good approximation to the thermal wind); then the warm air is on your right (in the northern hemisphere).

7.11 LATENT HEAT, HURRICANES AND TORNADOES

A great deal of weather is determined by the properties of water. Under normal atmospheric conditions water can, depending on temperature mainly, exist as a solid, a liquid or a vapour. No other material does that. What is important dynamically here is that a change of phase generally involves the emission or absorption of energy, and since this energy usually appears in the form of heat, we speak of the latent heat of fusion when we refer to the heat required to melt unit mass of something without changing its temperature, and the latent heat of vaporization as the heat required to vaporize unit mass of liquid without changing its temperature. Latent heat is the energy cost of getting the atoms or molecules into a more disordered state. The interesting fact about water is that it has a particularly high latent heat of vaporization, namely, some 2.5×10^6 Jkg^{-1}. (Compare ammonia, $1.2 \times 10^6 Jkg^{-1}$, propane $4.2 \times 10^5 Jkg^{-1}$, ethyl ether $3.9 \times 10^5 Jkg^{-1}$, carbon tetrachloride $2.2 \times 10^5 Jkg^{-1}$,). Consequently, when water evaporates it produces marked cooling, and when water vapour condenses it releases a substantial quantity of heat. Under certain circumstances that release of heat can be virtually explosive.

To get a feel for the size of the effect let us suppose we have some humid air containing, say, 1% by weight of water vapour. Suppose that water vapour suddenly condensed and formed a fog of water droplets. What would be the temperature rise of the air? We can calculate the amount of heat released very easily: it is just $2.5 \times 10^6 J$ for every kilogram of water vapour condensing, that is, $2.5 \times 10^6 J$ released in every 100kg of air. If this heat were entirely absorbed by the air the temperature rise can be calculated from a knowledge of the specific heat of the air, that is, the heat required to raise unit mass one degree Kelvin. A very rough, but easily remembered, figure for simple gases is $1 JK^{-1}g^{-1}$. Thus 100kg requires $10^5 J$ to rise in temperature by one degree, and hence the latent heat released by the condensing vapour, if none of it is lost, will raise the temperature by 25K, which is very substantial. Our calculation is very crude, but it illustrates the point that water vapour can act as a potent source of energy.

In the tropics, over the ocean, the humidity of the air in late summer and autumn, may reach many times the amount we have been considering, and such conditions can lead to the formation of a hurricane (another name for which is typhoon). Moist homogeneous air over warm ocean is extremely unstable. Hot air rising in thunder clouds causes air around the low pressure to converge, and spiral under the action of Coriolis force. If this spiralling action becomes sufficiently intense the pressure at the centre may drop so much that cold air from the stratosphere is sucked down and the tropopause (see Chapter 8) is brought down to ground level. As surrounding moist air spirals rapidly inwards towards the low pressure area it expands and cools adiabatically, and cools further as it rises around the eye. This double lowering of the temperature produced by the subsidence of stratospheric air and the adiabatic cooling, causes a massive condensation of water vapour once the air has risen, and the consequent large release of heat in a region tens of kilometres around the eye, retained by rain and air, fuels further evaporation from the sea and convergence of moist, expanding air. As the air rises, it cools. Water vapour condensing releases heat which drives a further rise and more convergence of air, spiralling inwards and upwards. The instability may grow eventually into a fully-fledged hurricane with a clear-weather 'eye' some 40km across, where air subsides from above, and this is surrounded by 'hurricane-force' winds carrying torrential rain. Over land, separated from its source of energy, the oceans, the hurricane soon weakens and disappears, but not before scything a swathe of destruction.

The fearsome energy residing in water vapour cannot be unleashed as easily over land as it can over sea. Nevertheless in the locality of developing thunder clouds, substantial up-draughts can cause surrounding air to converge and spiral inwards and upwards in much the same way as in the case of the hurricane. If that air is particularly humid, the condensation may power an instability which causes the formation of a narrow cloud which twists down to ground level in a circle roughly 100m across. Enormous radial pressure gradients build up, whose effect is a moving explosion which lasts for several minutes. Such is the tornado, the most graphic representation of the latent energy of water vapour. Though not unknown, tornadoes are rare in Britain. In central U.S.A. they are relatively common, with a maximum frequency of one per year per 80km square (Figure 14).

Tracking hurricanes has been made a good deal easier with the advent of weather satellites. Looking down on the continually evolving patterns of clouds over the surface of the earth with eyes which see in the infrared as well as in the visible they transmit electrical signals which can be converted into photographs. Two such photographs appear in Figure 7.15, and other examples have already appeared in Figure 7.11. Geological as well as weather patterns are very clearly shown. Long-term, as well as short-term trends can be investigated as Figure 7.16

Figure 7.14(a) (*above*) – Tornado seen from Cotton Exchange, Houston.
(Photo T. Anono and R. S. Scorer)
Figure 7.14(b) (*below*) Hurricane at 25°W, 50°N in process of becoming an extra-tropical cyclone (hence no eye), 15 Sept. 1978. (Photo from satellite NOAA/ESSA, by University of Dundee)

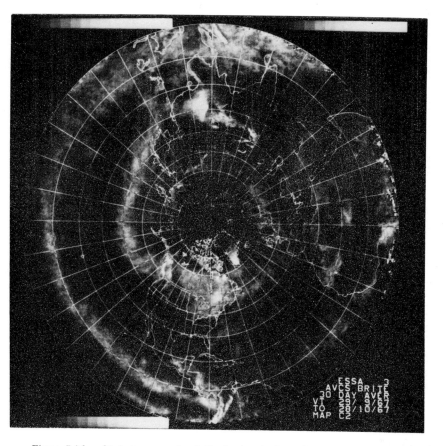

Figure 7.16 − 30-day average cloud distribution, built up from photographs by polar orbit satellite (NOAA/ESSA). The northern spiral indicates jet stream. Another cloud belt at 10°N is evident.

shows. Averaging the cloud cover seen by a satellite in a polar orbit (which is the orbit a satellite must have to cover the whole global surface) the circum-polar jet stream in the Northern Hemisphere is rendered strikingly visible. Also rendered visible is a persistent stream 10°N of the equator stretching across the Pacific.

7.12 OCEANIC CURRENTS

Currents in the oceans may be wind-driven or thermally-driven. In the latter case, evaporation of surface waters is the vital process. What happens is that evaporation, being essentially the escape of the hotter molecules, cools the water and at the same time increases the salinity, since salt remains in solution. Both effects cause the surface water to become denser than it was, and so it sinks and squeezes out laterally the deep, cold water which makes up the bulk of the ocean. Elsewhere, the deep cold layer gradually warms, becomes less dense, rises to the surface, and is replaced by a lateral inflow. The currents produced by these vertical movements are quite complex and rather slow. Evidence exists for deep, cold water remaining so in the ocean for 1,000 years. These thermo-saline currents are manifestations of the· ocean acting like the atmosphere does as a vast heat engine, though a more ponderous one than that of the atmosphere because of the much larger heat capacity of the ocean.

The much more potent currents of the ocean are driven by the friction between wind and surface. The Gulf Stream arises ultimately from the action of the prevailing westerlies in moderate latitudes and of the north-east trades in the tropics. It turns out however, that the direct action of the wind alone is quite insufficient to explain the substantial flow of water which is typical of a major oceanic current. How then do such currents arise?

All that the direct action of wind can do is to attempt to drive a thin surface layer of water along with it. If frictionless, this layer would move on average at right-angles to the wind by the influence of the Coriolis' force. But it is not frictionless. It attempts to drag along a thin layer immediately beneath it in a direction at some angle to the wind direction. The Coriolis' force also influences the motion of this second layer, which in turn tries to drag along a third layer.

The direction of motion therefore spirals from layer to layer, the magnitude of the motion getting weaker and weaker and after roughly some 100 metres, disappears. The total bulk of water affected by the wind in this way is known as the Ekman layer, after the Swedish oceanographer.

Although the direction of flow varies markedly with depth through the Ekman layer, the layer itself may be expected to behave overall roughly like a frictionless slab, since, by definition of the layer, the underlying water remains unaffected by its motion. Consequently, the average flow within the Ekman layer will be at right-angles to the wind direction. This flow is the direct result of wind action, but it is of insufficient magnitude to account for the observed currents of the oceans. What happens is that the flow in the Ekman layer piles up water, and the resulting pressure affects the whole depth of water, and not just the top 100 metres or so.

7.12.1 The Gulf Stream

In the North Atlantic, the westerlies force a southern flow and the north-east trades, a north-west flow. These surface currents literally pile up water in the Sargasso Sea, creating the analogue of an anticyclone, with a pressure-gradient acting throughout the depth of the ocean, forcing water outwards. To the north, this outward push gives rise to a deep current flowing eastwards; to the east the current flows southwards; to the south, it flows westwards. The westward boundary is complicated by the North American continent, which heavily modifies the simple anticyclonic flow by introducing friction: the pressure-gradient westwards cannot ensure a northerly flow. Furthermore, if the pressure-gradients in the north and south are of equal magnitude (though pointing in opposite directions), the latitudinal weakening of the Coriolis' force towards the equator means that the westward current to the south will have the greater velocity. The net effect is to build up pressure in the west. Thus the pressure-gradient acquires a west-east component which forces water south. The book-keeping effect of piling-up water in the Sargasso Sea is to produce a net southerly flow!

There has to be a northerly flow somewhere to balance this effect. But as we have seen, the combination of westward pressure-gradient and Coriolis' force cannot act effectively, because of the proximity of the continental coast-line of America. Where then is the northerly flow to come from? The answer is that the net component of flow to the south builds up a south-north pressure-gradient, which drives a direct northwards flowing current. The latter avoids being deflected eastwards by the Coriolis' force by exploiting the friction with the American coastline. It does this by being a fast flow rubbing against the continental shelf. This fast current is the Gulf Stream. It returns the water squeezed south over a large region of the north Atlantic, in a fast northerly flow along the western boundary. The feature of a fast, western-boundary current is common to most oceanic currents. (In the southern hemisphere, it would be directed southwards). (Figure 7.17).

The climate of Western Europe, insofar as it is influenced by the Gulf Stream, depends on a complicated chain, which begins (if a circle can have a beginning) with the production of the mid-latitude westerlies and of the tropical trade-winds by the atmospheric engine; leads to the pile-up of water in the Sargasso Sea by wind-driven Ekman layer currents, and the subsequent net southerly motion over most of the Atlantic; and ends with the rapid, northerly flow which exploits friction with America to avoid the effects of the Earth's rotation.

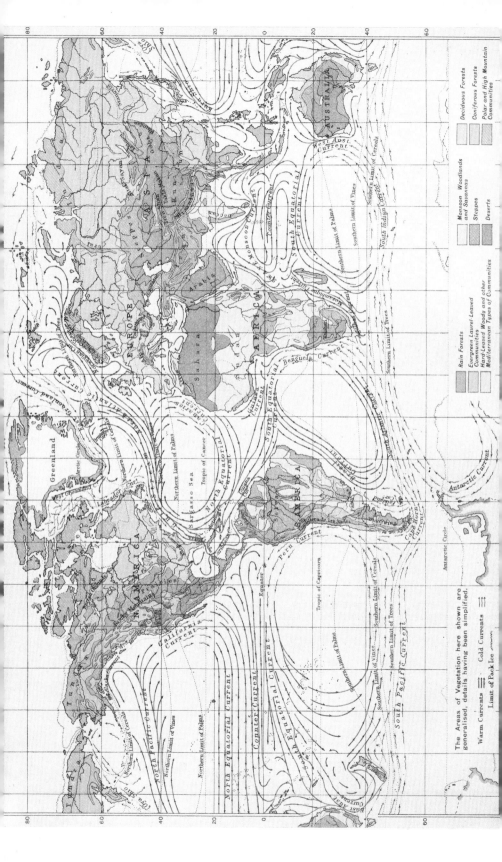

The Areas of Vegetation here shown are generalised, details having been simplified.

Warm Currents ≡≡≡ Cold Currents ⋙ Limit of Pack Ice

Rain Forests

Evergreen Laurel-Leaved Communities

Lee-ward Woody and other Mediterranean Types of Communities

Monsoon Woodlands and Savannas

Steppes

Deserts

Deciduous Forests

Coniferous Forests

Polar and High Mountain Communities

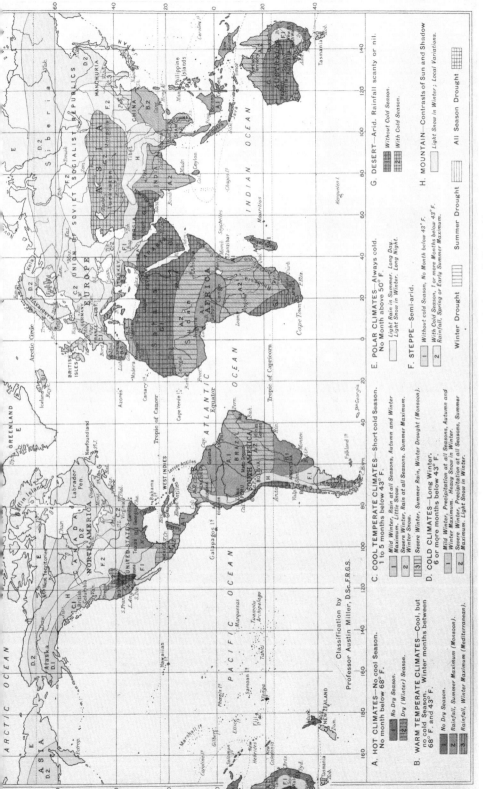

Figure 7.17(b) – Climatic Regions   (Copyright, John Bartholemew & Son Ltd.)

FURTHER READING

Allen, C. W., *Astrophysical Quantities,* (3rd ed., University of London, Athlone Press, 1973).

Dyson, F. J., 'Energy in the Universe (*New Frontiers in Astronomy*, Sept. 1971).

Fraser, *Understanding the Earth,* (Penguin, 1976).

Gregg, M., 'The Microstructure of the Ocean', (p. 64, *Scientific American,* Feb. '73).

Handbook of Physics in Chemistry, (Chemical Rubber Co., 1973).

Kelly, M., 'Physical Basis of Climatology', (p. 305, *New Frontiers in Astronomy,* July, 1976).

McIntosh, D. H. and Thom, A. S., *Essentials of Meteorology* (Wykeham Publication, London, 1969).

Ridley, B. K., *Time, Space and Things* (Penguin, 1976).

Scorer, R. S., 'Environmental Aerodynamics' (Ellis Horwood Ltd., Chichester, 1978).

Sutton, O. G., *Understanding Weather,* (Penguin, 3rd edition, 1969).

8

The Atmosphere

Sky – what a scowl of cloud
Till, near and far,
Ray on ray split the shroud
Splendid, a star!

Robert Browning (1812–1889)
The Two Poets of Croisic.

8.1 COMPOSITION

Much of the original atmosphere of the earth must have consisted of the two most abundant elements in the universe, hydrogen and helium. Helium, being an inert gas, could not chemically combine to become a component of a molecule heavy enough to be held by the earth's gravitational field, and has virtually all disappeared. Hydrogen, through its combination with oxygen, carbon and other elements, has been retained to a comparatively minor extent, but the vast proportion has been lost, and only the merest trace exists in the atmosphere as hydrogen.

The next most cosmically abundant elements are nitrogen, oxygen and neon. Neon, like helium, is an inert gas and cannot form heavy chemical compounds, not even diatomic molecules, as can nitrogen and oxygen. Nevertheless, its atomic weight is within a factor of two of the molecular weights of nitrogen and oxygen, and in fact greater than a molecule such as ammonia (NH_3) which might be expected to be fairly abundant, and also water (H_2O) which certainly is abundant. It is interesting therefore that it is present in the atmosphere only to the extent of about 0.002% by volume. If the earth's gravitational field is too weak to retain neon, how has it retained so much water, nitrogen and oxygen? This suggests that the composition of the atmosphere, and indeed the oceans, is not purely static but is determined by a dynamic balance. Unlike neon, nitrogen and oxygen undergo chemical reactions which prevents these elements escaping (Table 8.1).

Oxygen is an extremely reactive element. Nevertheless it makes up some 21% by volume of the atmosphere. Nitrogen, which is comparatively much less reactive, makes up about 78% by volume; argon, an inert gas actually heavier than neon but a product of crustal radioactivity, makes up about 1%; and carbon dioxide (CO_2) 0.03%. The water vapour content in the lower atmosphere varies between 0.1% and 3% by volume. That nitrogen composes the bulk of the atmosphere is perhaps not surprising in view of its cosmic abundance and the comparatively low reactivity of molecular nitrogen. That oxygen is so abundant is surprising, which again points to dynamic processes at work continually replenishing the oxygen that reacts with other chemicals.

Such a process is the photosynthesis of oxygen by plants according to the following equation:

$$CO_2 + H_2O \xrightarrow{\text{light}} \text{carbohydrates} + O_2$$

The presence of oxygen in the atmosphere is therefore a byproduct of vegetable life. Living cells generally appear to have evolved in an atmosphere that was a reducing one, rather than the oxidising one of today. The anaerobic bacteria, living at the present day in the muddy bottoms of lakes, cannot survive in air. Oxygen is a poison to them, and indeed to all living cells. Before plants evolved, the atmosphere was almost certainly a reducing one. Life had to adapt to neutralise, or better still, exploit the poisonous exhalation of oxygen. It has done so very successfully. So much so that most life cannot function without it. In the early days of the earth the atmosphere contained much more hydrogen than it contains now. Any oxygen exhaled by proto-life would have been rapidly removed as water. The gradual loss of hydrogen and the proliferation of anaerobic organisms exhaling oxygen inevitably changed the atmosphere to one which was oxygen rich.

Table 8.1 — Composition of the Atmosphere

Material	Percentage by volume in dry air
Nitrogen	78
Oxygen	21
Argon	0.93
Carbon Dioxide	0.03
(Water vapour)	(0.1–3.0)

Life processes also account for the abundance of nitrogen in the atmosphere. Although the problem is less urgent for nitrogen than it is for oxygen, because of the former's comparative inertness, it is nevertheless just as necessary to account for the fact that nitrogen is so abundant, because once molecular nitrogen reacts to form ammonium compounds and nitrates, nitrogen becomes a very reactive element indeed. Again, biological processes are at work. Plants take up nitrates in the soil, and death and decay returns nitrogen to the soil as ammonium compounds, which bacteria convert partly into molecular nitrogen and partly into nitrates. Animals get into the cycle by eating plants, and excreting, dying and decaying.

The composition of the atmosphere is therefore by no means a static thing. Exhalation and volcanic activity, chemical reaction and photosynthesis, the gaseous products of radioactivity, the factories of man, the loss to outer space — all of these processes combine, and have combined in the past to produce, and maintain, our thin shell of atmosphere.

8.2 STRUCTURE

The atmosphere exhibits a structure which varies with altitude (Figure 8.1). The various regions are denoted by *spheres*, and their tops are denoted by *pauses*.

One of the principal divisions is defined by composition. At altitudes up to 100km, the composition remains constant, with the chemical components well mixed. This region is known as the homosphere. Above 100km is the heterosphere, in which the molecular composition depends upon diffusion out of the homosphere. Another important region, defined in terms of physical composition, is the ionosphere which stretches up from 70km. In this region, atoms and molecules are ionized by the incoming solar radiation, and the resulting free electrons and charged ions make the ionosphere a conductor of electricity, enabling it to reflect radio waves. This process of ionization is one which protects life on earth from harmful x-ray and far-ultraviolet radiations. By the time it has reached the base of the ionosphere, all radiation of wavelength shorter than 1800 Å (photon energy greater than 6.9eV) has decreased in intensity by at least a factor of three from its extra-terrestrial value. The radiation decreases in intensity exponentially according to the law of optical absorption. The ability of atoms and molecules of oxygen and nitrogen in this wavelength range to absorb is extremely high, and practically none of the high-photon-energy radiation reaches the lower atmosphere.

But composition is not the only factor dividing the atmosphere — there is temperature. As everyone who has enjoyed the view from the top of a mountain knows, it is colder at high altitudes. The temperature falls by about 6.5K for every kilometre, on average. It is possible, locally, for a temperature inversion to occur, warm air riding over cold. Many meteorological conditions are known in which such inversions occur, but on a large scale, geographical and temporal, there exists a linear fall of temperature with height. This fall continues to a temperature of about 220K, which is reached at an altitude of about 10km near the poles and about 18km near the equator. The region up to this line is known as the troposphere, and the upper boundary is called the tropopause. The troposphere is the domain of weather and water vapour.

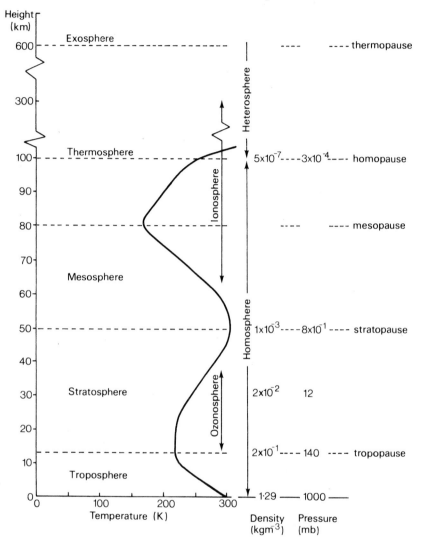

Figure 8.1 – The Atmosphere.

At the tropopause the temperature remains constant for between 10km and 20km and then begins to rise and reaches a maximum of 290K at an altitude of 50km. This altitude is known as the stratopause, and the region between the tropopause and stratopause is the stratosphere. Why does the temperature rise?

One would expect the temperature to continue to fall as one got further away from the ground which, absorbing all solar radiation which reaches it, is a warm 290K (on average). The answer is associated with the existence of a layer of ozone (O_3) in the stratosphere. The ozone layer lies between 12 and 50km, having a maximum concentration at about 25km of about ten times the sea-level value, i.e. only about 0.00001% by volume. Small though the concentration is, its presence is important because it absorbs ultraviolet light in the wavelength range 1800 to 3,000 Å, just the range of short wavelength radiation not absorbed in the ionosphere. As a consequence, it gets hot, and since most of the radiation is at the top, the temperature rises with altitude through the stratosphere. The region between about 25km and 50km in which the temperature rises is known as the ozonosphere. The concentration of ozone in, and the width of the ozone layer, is delicately dependent on the radiation intensity, temperature and pressure. Ozone is produced by the action of ultraviolet radiation on oxygen, and its decomposition back into oxygen is temperature and pressure dependent. Since its temperature depends on the absorption of radiation, which in turn depends upon the concentration, all these quantities are intimately related. The ozonosphere comprises the small layer of the atmosphere where this delicate physico-chemical balance between the creation and destruction of ozone determines the physical conditions. Above 50km ozone is too dilute.

With decreasing absorption of radiation immediately above the stratopause the temperature-curve sluggishly reverses direction, and the temperature falls with altitude to about 200K at a height of 80km. Above 80km it abruptly reverses its trend. The region between the stratopause and the so-called mesopause at 80km is known as the mesophere. Above the mesosphere is the thermosphere, in which the temperature increases rapidly with altitude owing to the intense absorption of the ionising, short-wavelength radiation.

Above about 600km the atmosphere is so rarified that collisions between atoms occur too infrequently for there to be a meaningful temperature. This region is known as the exosphere. Above about 1,000km the atoms are on average half-ionised; and above 1,200km hydrogen, which gradually increases its relative concentration above the homopause, becomes the dominant constituent of our atmosphere.

Placed even further outwards are the radiation, or Van Allen, belts. The inner is at $1.6R$, where R is the radius of the earth (about 6,000km), and the outer is at 3.5R. They consist of charged particles, principally charged nuclei emitted by the sun, captured by the earth's magnetic field. The charged particles spiral around the field lines from one pole to the other. Charged particles which escape the magnetic field trap and enter the atmosphere are responsible for the glow of the night sky and the aurorae, which originate in the emission of light by excited atoms (invisible during the day) some 100 to 1,000km above the surface.

8.3 PRESSURE AND DENSITY

The total mass of the atmosphere is 5.30×10^{18} kg. Because it is pulled downwards by the earth's gravity, it is far from uniform in density, being concentrated at ground level. That it extends so far up in altitude is because the pressure opposes the gravitational pull. More accurately, it is the vertical pressure-gradient force (see Section 7.8) which stops all the atmosphere falling on to the earth's surface. A unit volume of the atmosphere therefore experiences two forces: the pressure-gradient force pointing upwards, and the gravitational force pulling downwards (Figure 8.2). For no net vertical movement of the atmosphere these forces must cancel each other out at all heights. It turns out that, to a good approximation, this happens if the pressure and density of the atmosphere decreases exponentially upwards, and that is what is observed to be the case. At the surface the pressure is around 1000 millibars (equal to about 1kg-weight per square centimetre), decreasing by a factor of 2.7 (the ubiquitous base of natural logarithms, e) every 7km. This distance of 7km is known as the **scale height**. Density also decreases exponentially from its surface value of 1.29kg m^{-3}, with the same scale height.

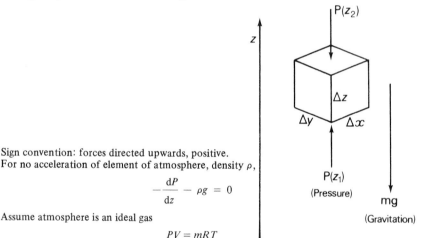

Sign convention: forces directed upwards, positive.
For no acceleration of element of atmosphere, density ρ,

$$-\frac{dP}{dz} - \rho g = 0$$

Assume atmosphere is an ideal gas

$$PV = mRT$$

where m = mass, R = gas constant per unit mass, V = volume, T = absolute temperature.

$$\therefore \quad P \quad = \quad \rho RT$$

$$\therefore \quad \frac{dP}{dz} \quad = \quad -\frac{g}{RT} P$$

If T constant, solution is $P \quad = \quad P_0 e^{-z/H}$

where H (the scale height) $\quad = \quad \dfrac{RT}{g} \approx 7\text{km}.$

Figure 8.2 – Variation of Pressure with Height. (In practice H varies with height and is a measure of the variation of temperature)

8.4 ABSORPTION AND SCATTERING

Of the solar radiation incident on the atmosphere only a fraction survives the absorption and scattering processes which occur in the air, and reaches the earth's surface. Because the sun, moon and stars can be seen reasonably clearly in the absence of cloud we can conclude that, at least in the visible region of the spectrum, this fraction is appreciable. Indeed, that we can see appreciable distances on the earth's surface on a clear day shows the atmosphere to be remarkably transparent to visible light. In other parts of the spectrum this is not the case, as we have already seen in the case of x-rays and ultraviolet radiation – the uppermost layers absorb wavelengths below 1,800 Å, and ozone layer absorbs wavelengths up to 3,000 Å. And in these layers, where absorption is comparatively intense, the temperature is comparatively high.

Absorption of electromagnetic energy ultimately produces heat inside the absorber. The absorption of a photon by an atom raises one of the atom's electrons to an excited state within the atom, or to the free-electron states outside the atom. In the first case, the atom is excited and may de-excite itself in several ways. Some of these ways, such as by a collision with another atom, or by emitting a photon which is absorbed by another atom, transfers the energy of the original photon to the rest of the gas. Once it has become shared out between a large number of atoms (or molecules) we can say that it has increased the temperature of the assembly. The sharing process is quite complex, and it takes time, but the end effect is that absorption leads to heating. A photon disappears and eventually the neighbouring material is heated by its energy.

If the electron is knocked completely out of the atom the process is called photo-ionization. While free, the electron may be buffeted by other photons and pick up kinetic energy from them. This process, involving the collision of a photon with an electron, is known as the Compton effect. Eventually the electron will recombine with an ion and energy will be released in the form of photons, or the recoil of neighbouring atoms. Again, the original photon's energy, plus any energy transferred from other photons to the free electron via the Compton effect, eventually ends up, wholly or partly, converted into heat – (partly, if photons are produced in the recombination or de-excitation process which happen not to be absorbed by the surrounding medium).

Atomic absorption generally involves photons of energy greater than, very roughly, 1eV. The ionization energies of atmospheric atoms and molecules range between 12eV and 15eV (12eV correponds to a wavelength of about 1,000 Å). Photons with less than the ionization energy can be absorbed only if a suitable excited electronic state exists exactly at the photon energy above ground-state. As it happens, none of the common air molecules can absorb visible radiation because they lack excited states at just these energies. Because the binding energies of molecular oxygen and nitrogen are about 5eV and 10eV respectively, the absorption of far-ultraviolet and x-ray photons can dissociate these molecules into their constituent atoms. Molecules (excepting ozone) therefore become increasingly rare with increasing altitude.

8.4.1 The atmospheric greenhouse

There is an abundance of absorption processes in the atmosphere for cutting off short-wavelength radiation (Figure 8.3). Ozone absorbs some of the visible light, but apart from this the atmosphere is transparent to these wavelengths. In the infrared, however, strong absorption bands can arise from the excitation of molecular vibrations. Molecular, rather than atomic, processes therefore dominate at these longer wavelengths. Water vapour absorbs various bands of wavelengths longer than 0.7 μm (7,000 Å) and carbon dioxide has strong bands at 13 μm and 17 μm. Curiously, there exists a 'window' around 10 μm where the absorption is small. This is just the wavelength at the peak of the emission of the earth. (It also encompasses the 10.6 μm and 9.6 μm radiation emitted by carbon-dioxide lasers, a fact exploitable for communication purposes). In spite of this window, the atmospheric molecular absorption in the near and far infrared means that practically all the thermal radiation emitted by the earth, and the infrared part of the solar radiation, is absorbed by the lower levels of the atmosphere which is comparatively rich in water vapour and carbon-dioxide, and to some extent ozone.

The atmosphere, principally through its water vapour and carbon-dioxide content, therefore acts like a greenhouse. Glass allows sunlight to enter and be absorbed, but it stops the emission of infrared radiation from the heated interior to escape directly. In absorbing the infrared light, it becomes heated and radiates infrared itself, some back into the greenhouse, some to the outside. This effect contributes to the higher temperature inside the greenhouse (though, as we mentioned in Section 5.8, the prevention of air motion is the most important in keeping greenhouses warm). In an exactly similar way the atmosphere acts radiatively like the glass in the greenhouse, allowing sunlight to penetrate, but blocking the earth's thermal radiation. The size of the effect depends upon the concentration of water vapour and carbon-dioxide present. Some concern has been expressed that man's industrial activity may be increasing the average content of carbon-dioxide and so accentuating the greenhouse effect (See Section 8.6).

As well as absorption on the atomic and molecular level there is absorption caused by dust particles.

Dust is carried into the atmosphere by the action of wind, industrial activity, and, explosively, by volcanoes. If carried into the lower stratosphere dust may remain there for months and even years. At higher altitudes layers are found which may be of extra-terrestrial origin, constituting the debris of matter picked up by the earth in its journey through space. In the troposphere clouds, which cover on average 50% of the earth's surface, contribute significantly to the reflection and scattering of solar radiation, but not to the absorption.

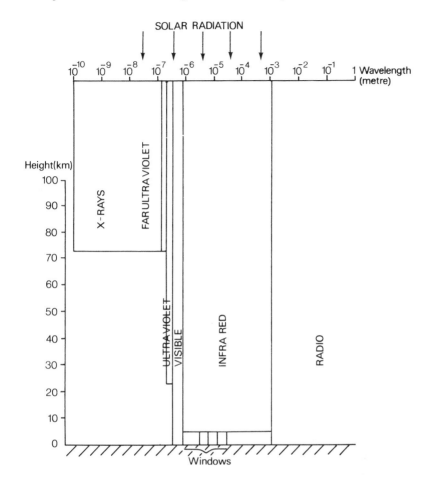

Figure 8.3 – Absorption of Atmosphere

8.4.2 The sky is blue

Besides absorbing radiation the atoms, molecules and particles of the atmosphere can scatter. In scattering processes the energy of the radiation is absorbed by the scatterer and immediately re-emitted, generally in a different direction. Unlike absorption, scattering involves little or no energy exchange. The principle mechanism is called Rayleigh scattering, and is characterised by its rapid dependence upon wavelength, shorter wavelengths being scattered more intensely than longer ones. The strength of scattering S, of an atom, molecule or small particle, is inversely proportional to the fourth power of the wavelength viz: $S \propto \lambda^{-4}$.

This rapid variation with wavelength comes about in the following way. If an electron is accelerated, it radiates. In classical electromagnetism all radiation derives from accelerated charges. Moreover the electric field-strength in the radiated wave is proportional to the magnitude of the acceleration. Because radiation intensity is proportional to the square of the electric field, intensity is therefore proportional to the square of the acceleration. Now, if a wave is incident on an atom it can be thought of as driving the electron into oscillation at the frequency of the wave. In a simple harmonic oscillation the acceleration is proportional to the square of the frequency. Consequently the oscillating electron will radiate a wave of the same frequency as the incident wave but with an intensity which is proportional to the fourth power of the frequency; or, since frequency is inversely proportional to wavelength, the re-emitted intensity is inversely proportional to the fourth power of the wavelength.

One of the consequences of scattering is that much of the solar radiation incident on the atmosphere is scattered back into space. Clouds are particularly good at this, and they are also good at diffusing sunlight. Another consequence is that part of solar radiation reaches the ground after many scattering events, and therefore arrives from all directions. The favouring of short wavelengths means that much of the visible scattered light will be blue, which is why the sky is usually that colour and not the black of space. It also means that objects viewed through a great thickness of the atmosphere, such as the setting or rising sun or moon, have a reddish tinge, the blue light being preferentially scattered out of the direct path. In a dense atmosphere, such as pertains on Venus, Rayleigh scattering would seriously inhibit the perception of objects unless they were close by. Also the distinction between night and day would not be so marked. Our terrestrial atmosphere is dense enough to give us blue skies, and twilight for a short period after the sun has set, but it is not so dense as to inhibit perception (at least on clear days) or to blind us to the splendour of the night sky.

8.5 THE RADIATION BALANCE

The atmosphere acts in quite a complex way to determine what happens to the radiation coming from the sun and that coming from the earth. Let us for a moment forget about details and look at the overall picture. The sun emits radiant energy uniformly in all directions. Part of this radiation falls upon the earth's sphere, situated about 150 million kilometres away. Some of it is reflected, some absorbed (whether by the atmosphere or the earth itself is at the moment of no concern). The absorbed energy heats the globe, and the globe radiates electromagnetic waves into space, approximately uniformly in all directions (if we ignore latitude differences of temperature). If the influx of energy from the sun equals the outflux of energy radiated by the earth, the terrestrial temperature assumes an appropriate steady value. Indeed, a steady average temperature implies that such a dynamic balance must exist. Whatever the details of the radiative processes within the atmosphere and at the earth's surface, a basic average temperature will be defined by equating energy input to energy output. This can be easily calculated (Figure 8.4). It turns out to be about 250K (about $-22°C$) which is obtained by taking the effective temperature of the sun to be 5800K, the fraction absorbed 0.65, the radius of the sun 6.96×10^8m and the earth-sun sun distance 1.50×10^{11}m.

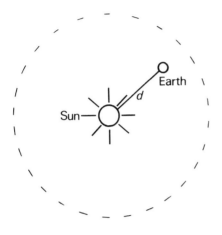

Intensity of radiation emitted by sun at sun's surface $= e_s \sigma T_s^4$
(e_s = emissivity, σ = Stefan's constant, T_s = absolute temperature)

Intensity at earth $= e_s \sigma T_s^4 \dfrac{R_s^2}{d^2}$

(R_s = radius of sun, d = sun-earth distance).

Energy per second received by earth $= e_s \sigma T_s^4 \dfrac{R_s^2}{d^2} \Pi R_e^2$

(R_e = earth's radius). Let f_e be fraction absorbed.

Energy per second absorbed by earth $= f_e e_s \sigma T_s^2 \dfrac{R_s^2}{d^2} \Pi R_e^2$

Energy per second emitted by earth $= e_e \sigma T_e^4 . 4 \Pi R_e^2$
(e_e = emissivity of earth, T_e = absolute temperature of earth)

At steady state, energy absorbed per second = energy emitted per second

$$\therefore e_e \sigma T_e^4 4 \Pi R_e^2 = f_e e_s \sigma T_s^4 \frac{R_s^2}{d^2} \Pi R_e^2$$

$$\therefore T_e = T_s \left[\frac{f_e e_s}{e_e} \frac{R_s^2}{4d^2} \right]^{1/4} \approx 250 \text{ K}.$$

Figure 8.4 – Large-scale radiation balance

Now, that temperature of 250K refers to a surface of the earth radiating freely into space. It therefore refers more nearly to the atmosphere rather than the earth's surface, whose radiation in the infrared is mainly absorbed by the atmosphere. This blanket effect of the atmosphere keeps the earth's surface significantly warmer than 250K. To estimate how much warmer, we can consider the atmosphere to be at 250K and radiating outwards into space and also inwards to the earth's surface, whereas the earth's surface radiates only upwards. If the earth's temperature is T, then the radiation balance between ground and atmosphere, and Stefan's Law of radiation, indicates that $T^4 = 2(250)^4$, and therefore, $T = 297K$, which is about right. The difference between 250K and 297K is a measure of the greenhouse effect.

The radiation intensities involved in defining an average temperature are for the whole electromagnetic spectrum. Let us now look at the energy balance in more detail by taking into account the factors mentioned in Section 8.4 and splitting the earth into two components: the atmosphere and the surface. Each of these interacts with the incoming solar beam, which is predominantly short wavelength radiation, and with the thermal radiation, which consists predominantly of long wavelengths. We therefore consider two components of the total radiation: solar radiation and the terrestrial beam. The energy exchange between the atmosphere and the surface is not entirely radiative. Air which is warmed by contact with the surface rises, and transports heat upwards by convection. Also, evaporation of water from the oceans cools the surface, and when water droplets condense, heat (latent heat of condensation) is passed into the atmosphere. Thermal conduction also transports heat from the surface into the air, though to a smaller extent than convection and evaporation. We must include the contribution of conduction, convection and evaporation in the energy exchange between the atmosphere and the surface.

Figure 8.5. illustrates the various processes involved, and the rough magnitudes of their associated energy fluxes. For every 100 units of solar radiation incident upon the upper atmosphere, only 22 reach the surface directly and are absorbed. Of the rest, 35 are reflected back into space (mostly from within the troposphere) 21 are absorbed by the atmosphere (mostly above the troposphere), and the remainder reaches the surface after scattering and diffusing through clouds. The earth radiates some 118 units, only 11 of which escape directly; the rest being absorbed by the atmosphere. The latter radiates 158 units, 54 of which travel into space and the rest return to earth. The outgoing thermal radiation (11 units directly from the surface and 54 from the atmosphere) therfore just balance the 65 units of solar energy which avoid reflection and become absorbed. Radiation processes alone, however, leave the atmosphere in debit by 30 units and the surface in credit by the same amount. A net balance is achieved via the processes of convection and evaporation, and a little conduction.

These quantities may be put on an absolute footing by equating the 100 units of incident solar radiation with the actual intensity, measured outside the atmosphere. This important quantity is known as the solar constant and has the value 1.35kWm^{-2}. (More accurately it should be known as the solar parameter, because it is not strictly constant.) How this energy is spread over the wavelengths was shown in Figure 7.4 (page 110).

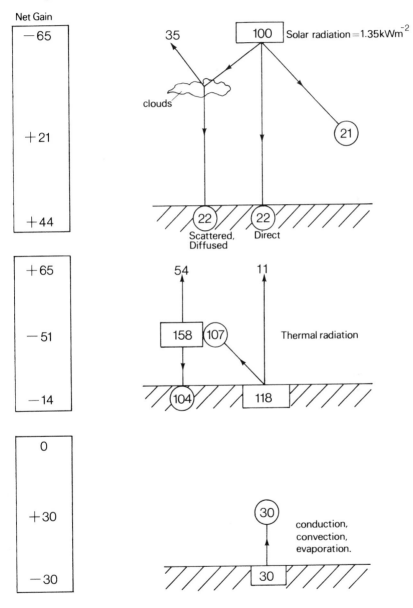

Figure 8.5 — Detailed radiation balance.

Let us emphasize that the numbers in Figure 8.5 are averages. Details of the radiation balance naturally vary with geographical location, time of day, latitude, season and water vapour content of the atmosphere. Time of day, latitude and season are particularly important since together they determine the angle at which a region of the earth presents itself to the solar beam through the day — the more acute the angle the more the surface area a given flux of radiation covers, and consequently the lower the effective intensity. For the same reason, north-facing slopes differ from south-facing slopes. Reflectivity, or albedo (see Section 5.8), is also a considerable factor — sand-dunes reflect more than forest — and the emissivities of the various features inhabiting the earth's surface differ appreciably. Considerable local and temporal departures from the global picture of radiation balance therefore occur.

8.6 POLLUTION

We rely on the warmth of the sun and the purity of the air for our survival. Yet many of our activities on the planet threaten to affect these vital elements adversely, and in some places at certain times, when London fogs or Los Angeles smogs occur, that threat has become a fact. On a global scale, such is the immensity of the atmosphere and such are the checks and balances built into that vast system, that any activity we indulge in is likely to have rather puny direct effects. Nevertheless, given the evident danger of many of the by-products of our technology, given the accelerating rate of global industrial activity, and given the possibility of long-term building up of pollutants, it is essential that an awareness of this problem permeate the whole consciousness of mankind.

There are four types of pollution: (a) **particulate**, (b) **chemical**, (c) **radiation** and (d) **thermal**.

8.6.1 Particles

Particles of one sort and another are the most obvious sort of pollutant, for example, where they exist as smoke, or industrial haze. They cut off sunlight, reduce visibility, and in large concentration irritate the respiratory system. We have already mentioned their role in determining the radiation balance. Fortunately man-made pollutants are mostly confined to the troposphere and are quickly washed out. Notice how clear the air is, after it has rained. Dust which gets into the stratosphere, on the other hand, disappears only very slowly. That fact makes dust emanating from volcanoes and desert sand-storms far more important than industrial particulate pollution.

8.6.2 Chemicals

The principle pollutants are carbon dioxide CO_2, (already mentioned in relation to the greenhouse effect); carbon monoxide, CO; sulphur oxides SO_2, SO_3 and sulphates (respiratory irritants); nitrogen oxides, NO, NO_2 (respiratory irritants); hydrocarbons (some of which are carcinogens); many other potentially dangerous materials such as lead, beryllium, mercury and asbestos. In addition, under the action of sunlight, the nitrogen oxides and hydrocarbons can produce the poisonous gas ozone, O_3, and other so-called photochemical oxidants. Moreover, the fear that chemicals in every day use may percolate upwards through the atmosphere and destroy the ozone layer has been expressed, and has even triggered legislation in the United States, Canada and Sweden. How well-founded such fears are is difficult to judge. Fears about long-term effects of chemical pollution are equally difficult to evaluate.

There is no doubt that the consumption of fossil fuels releases carbon-dioxide. If that consumption maintains its present rate, it has been calculated that the concentration of CO_2 in the atmosphere will be doubled in 200 years; and if the consumption continues to accelerate at the present level, that time will be cut to a mere 50 years. Of course, such estimates are bound to be crude. The difficulties of allowing for the stabilizing effects inherent in the atmosphere and biosphere – (the homeostasis of Gaia, the living earth, Chapter 10) makes calculations of this sort very hazardous. Nevertheless, the qualitative trend of increasing concentration of CO_2 in the atmosphere cannot seriously be doubted, nor can this be regarded without concern. Measurements show that CO_2 content oscillates seasonally. During the growth season, plants use up CO_2 in photosynthesis and release it during autumn and winter. But superimposed is an increase of CO_2 which amounted to some 5 per cent over the period 1958 to 1976. Without the stabilizing effect of cloud cover, a continuing increase would lead, via the greenhouse effect, to a rise in temperature of the earth's surface, and to subsequent melting of the Arctic and Antarctic ice sheets; and to an accompanying rise in sea-level, which would drive coastal communities inland. On the other hand, industrial consumption of fossil fuels also produces a good deal of smoke and some particles remain in the atmosphere, cutting off solar radiation and thereby producing a cooling effect. Which process is winning? It is hard to say. The evidence is that since 1940 the northern hemisphere has cooled very slightly, but this may be part of a naturally occurring cycle and owe nothing to man's activities. The fact is, we do not know. But we do know that polluting the atmosphere with CO_2 is potentially dangerous. How dangerous is a matter of opinion at the present time.

8.6.3 Radiation

There is no doubt about the lethal properties of intense radiation. It is far more difficult to apprehend the level of danger to life of prolonged exposure to weak radioactivity which, by killing micro-organisms, may be marginally beneficial! Nevertheless it is usually assumed that any radiation is bad, and extremely elaborate precautions are taken to ensure that radiation does not become a pollutant. The enormously long life of some of the radioactive ash produced in nuclear processes (up to 2.3 million years in the case of caesium-135) makes safeguarding against pollution a major task. The topic of radioactivity is treated at greater length in Section 8.9.

8.6.4 Heat

In principle all the particulate, the chemical, and the radiation pollution produced by our technology could be contained, and none allowed to escape into the atmosphere. By the very nature of our industrial activity, and indeed of any activity whatsoever, the production of heat cannot be prevented and its eventual escape into the environment is not ultimately preventable. If this heat arises from any source other than from solar or geothermal energy, directly or indirectly, it may be considered a pollutant. All our industrial processes require energy. If this energy is obtained from fossil fuels or nuclear processes, such energy arises from reactions which inevitably disturb the thermodynamic equilibrium of the globe. These reactions release energy, and at a far greater rate than would otherwise occur naturally. Direct tapping of solar or geothermal energy does not create new energy but merely diverts what already exists into useful channels. Thus the generation of energy from solar cells, hydroelectric plants, tides, hot springs, etc. does not alter the energy balance of the planet to a first approximation, whereas burning coal, splitting uranium or fusing hydrogen are processes which do. Most of the energy released is heat, such are the typical efficiencies of conversion.

At present thermal pollution is confined to local effects. If the production of energy reached 1% of the input from the sun, a simple calculation using the blackbody radiation law shows that the average temperature of the earth would rise about $0.7°C$ ($1°F$). The rate of solar energy incident on the earth is 1.72×10^{11} MW (megawatts), of which about 1.1×10^{11} MW is absorbed by the atmosphere and by the surface. Thus an energy production rate of about 1.1×10^{9} MW is at the level corresponding to 1% of the solar input i.e. at the level when appreciable global changes occur. Present global production of energy by all methods is roughly a factor 100 less than this (see Section 8.10).

8.7 ATMOSPHERIC ELECTRICITY

It is one thing to live in a greenhouse, and quite another to find that one is also living inside an electrical capacitor. The positive plate is the ionosphere (a rather diffuse plate) and the negative plate is the earth itself, and the system is nothing less, electrically, then a gigantic spherical condenser, with essentially the lower (20km) atmosphere acting as a rather leaky dielectric with peculiar properties. The condenser is charged – the surface negative charge on the earth is 1.2×10^{-3} Ckm^{-2} (coulombs per sq. km.) and since the atmosphere conducts electricity, because of the presence of ions, this charge continually leaks away. On a fine day, a steady discharge current flows through the atmosphere, transporting positive charge down from the ionosphere and negative charge up from the ground. Over the whole earth it is estimated that this current amounts to some 1,500 A (amperes). Since the electrical resistance of the atmosphere between the whole earth surface and the beginning of the conducting layer at 20km is some 200Ω (ohms), this implies a steady voltage difference across some 20km of lower atmosphere of 3×10^5 V (volts), which in turn implies an average electric field (pointing downwards) at 15Vm^{-1}. If we measure this field, however, we find that it is by no means uniform with altitude, being 130Vm^{-1} at the surface and decreasing rapidly with height. However complex its variation, its principle feature is that it exists as an invisible but real part of our physical environment.

Two questions immediately arise. If a discharge current continually flows how is it that the charge has not been neutralised long ago? The discharge of a condenser of capacity C, through a resistance R, follows a well-known exponential law. Where e is the base of natural logs, the charge decays to e^{-1} of its initial value in a time given by RC. The capacitance of the earth is roughly 6.8×10^{-2} F, (farad.) and so the discharge time constant is of the order of 10s. Thus in a matter of minutes all the charge should have disappeared. Clearly, processes are at work which replace the charge as fast as it disappears. What are they? The second question is, why is the polarity such that the earth is negative and the ionosphere positive? There are no complete answers to these questions, but it is generally accepted that the role of thunderstorms is a vital one, and at base, the whole problem of the electrification of the atmosphere is closely tied up with the electrical properties of ice and water.

At any one moment, on average, there are some 2,000 thunderstorms raging somewhere in the world (Figure 8.6). Usually, thunderclouds tower upwards from 3 to 10km in moderate latitudes, and up to twice as high in the tropics, reaching above the altitude at which water freezes (Figure 8.7). Strong up-currents within the cloud carry water-droplets upwards to an altitude where they freeze, amalgamate, and fall as hail. Various electrification processes can be associated with this movement, involving the ways in which water droplets and ice may have become charged — (the question of how cloud electrification occurs is a very lively topic which we will discuss later) — and a characteristic charge distribution builds up in a matter of 10 to 20 minutes in various regions of the cloud at various times, such that the base of the cloud becomes negatively charged, and the upper regions positively charged. Consequently, the fields below and above the cloud are opposite in direction to the fine-weather field. At ground level the strength of the field may reach the breakdown value for damp air, which is about 10^6 Vm^{-1}, and a cloud-to-ground lightning flash occurs. Fields less than the breakdown strength are concentrated and enhanced at points and sharp edges, such as are provided by grass and trees, and a local corona discharge, so called because the point or edge is crowned with a visible halo, can take place vigorously. Whether the current occurs via an explosive lightning stroke or via the more gentle point discharge, the net charging effect is the same:

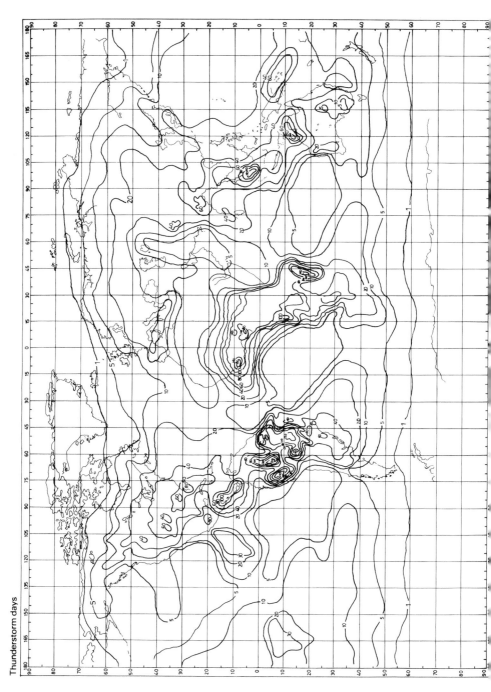

Thunderstorm days

Figure 8.6(a) −
World thunderstorm frequency. The lines join places with equal numbers of thunderstorm days a year.
(World Meteorological Organization Publication W.M.O./OMM 21TP21, 1956).

negative charge is transferred from cloud to ground. Above the thundercloud, the field transports positive charge to the ionosphere. Thus thunderstorms are the charge-generators, which maintain the charge, and the polarity, on the terrestrial capacitor.

Figure 8.6(b) – Line drawing of thunderstorm frequency in the British Isles 1931–1960. Crown Copyright reproduced with permission of the Controller of Her Majesty's Stationery Office.

8.7.1 Charge generation

But how does the charge generation work? Once more the answer lies in properties of water, in particular, the property of charge-transfer. Within ice and water, charge can move from one place to another *via* several mechanisms, and this facility allows the possibility of separating positive and negative components. Such a separation may occur in several ways. The most basic is via the thermo-electric effect. If two pieces of ice are put in contact, the warmer piece will generally acquire a negative charge, and the colder a positive charge. The collision of ice particles with different temperatures in a cloud will therefore result in the particles becoming oppositely charged during the period they are in contact. Charging ice by friction may also occur, and often the smaller particle rubbing against the larger acquires the negative charge, perhaps through the thermoelectric effect, because it gets slightly hotter. Once an electron field is set up, water droplets become electrically polarized, and if they splinter, opposite charges may be physically separated. In this mechanism the temperature gradient is replaced by a voltage gradient. Other charging processes are thought to occur in connection with the freezing of water, evaporation, and the melting of ice.

Once the charges are separated, say by a collision of ice particles, what is to stop them recombining? After all, electric forces are exceedingly strong. The answer is that there has to be enough turbulence and gravitational separation to overcome this stabilizing effect. Roughly, the kinetic energy of the air motion has to exceed the electric field energy. An updraught of over 10km hr^{-1} would be sufficient to separate oppositely charged particles (Figure 8.7). and allow gravitational separation to build up a vertical field. A simple picture of charge generation within a thunderstorm is one in which hailstones, heavy enough to fall through the updraught and warmed by the latent heat of super-cooled droplets freezing on them, acquire negative charges through collisions with small ice particles borne aloft. In this way the base of the cloud could acquire a negative charge and the top, a positive charge within a matter of 10 to 30 minutes, sufficient to cause a lightning discharge either within the cloud, between clouds, or between cloud and earth.

Estimates of charging rate associated with lightning and point-discharge actually exceed the rate required to counteract fine-weather conduction. Since fine-weather and thunderstorms are not the only alternatives open to the weather, the discrepancy is not surprising. The balance, indeed, is established by the bad-weather discharge current, which is associated with the fact that rain and snow are generally positively charged. Table 8.1 summarises Hutchinson's estimates of the charge per square kilometre added to the earth every year by the four processes.

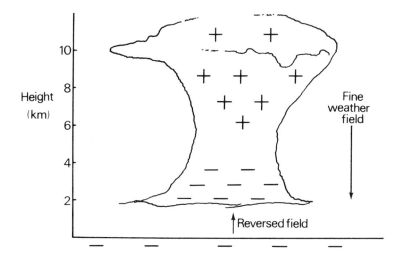

Figure 8.7 – Simplified charge distribution in thundercloud. Re-combination of charge is prevented provided the kinetic energy of turbulence exceeds the electric field energy, viz $\frac{1}{2}\rho v^2 > \frac{1}{2}\varepsilon E^2$ (energy densities) (ρ = density, v = velocity, ε = permittivity, E = electric field). For example if E = 10^6 Vm^{-1}, then with $\varepsilon \approx 10^{-11}$ Fm^{-1}, $\rho \approx 1$ kgm^{-3}. v > 3 ms$^{-1} \approx 10$ km hr^{-1}.

Table 8.2 – Strength of Charging Processes

Process	Contribution ($Ckm^{-2} yr^{-1}$)
Fine-weather conduction	+90
Rain and snow	+30
Lightning	−20
Point-discharge	−100

8.8. IONS IN THE ATMOSPHERE

The ions responsible for the fine-weather conduction at ground level cannot arise, as they do in the upper atmosphere, from the action of the ultraviolet and x-ray components of the solar radiations, because these do not reach the earth's surface. They are caused instead by the α-particles, β-particles and γ-rays emitted by radioactive atoms in the earth's crust, and in the lower atmosphere itself, and, to a lesser extent, by cosmic rays. Ionization by radioactivity occurs principally on or near land and scarcely at all in mid-ocean, whereas that caused by cosmic rays is the same over land and ocean. In spite of this, the longer life of ions over oceans make the steady state ionic concentration (about 4 to 5×10^8 m^{-3}) the same for land and sea.

Without cosmic rays the atmosphere over the oceans would be significantly less conducting. These rays are of extra-terrestrial origin. Some derive from the sun, but the rest appear to have a truly cosmic origin in that their intensity does not at all depend upon direction. Outside of the earth's atmosphere they consist of energetic nuclei, mostly protons, many of which have gigantic energies, far beyond what can be produced by the man-made accelerators of high-energy physics. Smashing into the nuclei of the atmospheric atoms, they initiate a shower of elementary particles travelling close to the speed of light, some of which reach ground level, and become a factor in the mutation of genetic material, as well as ionizing the air.

It might be expected that the chemical composition of the ion population would reflect that of the atmosphere, and that oxygen molecule ions and nitrogen molecule ions would dominate. In fact, the collision processes which occur in the lower atmosphere are so rapid — one every tenth of a nanosecond, for each particle roughly — that the initial ions which are produced are rapidly transformed by charge-transfer processes and chemical reactions. Of the products, the ions of most importance to conduction are the so-called fast ions, which besides being light and highly mobile are most numerous. The positively-charged fast ion is the hydrated hydroxonium ion, which has the composition $(H_3O)^+$ $(H_2O)_n$, where n in the troposphere is about 6 or 7, but varies with water-vapour content. There are two negatively-charged fast ions, the hydrated oxygen molecule ion with the composition $O_2^-(H_2O)_n$, and the hydrated orthocarbonate ion $CO_4^-(H_2O)_n$. In an electric field of 1 Vcm^{-1} such ions would drift with speeds of about 1cms^{-1}, but their mobility obviously depends upon the number of water molecules which are attached to the ion. The more humid the atmosphere, the slower the ions move.

Not surprisingly, therefore, the conductivity of the air varies with time. Even in the absence of changes in humidity the conductivity is constantly undergoing fluctuations because ionization is dependent on the arrival of an ionizing particle. In between such events the conductivity decays by as much as a factor of 3. It is also observed that there are diurnal and annual variations, and in that respect the atmospheric electric battery partakes of the cyclical behaviour of the atmosphere in its other activities.

8.9 THE RADIOACTIVE ENVIRONMENT

The continual production of ions in the atmosphere is a manifestation of another invisible part of our environment, the flux of atomic and sub-atomic particles associated with cosmic rays and radioactivity. This type of radiation, when it is intense, is known quite certainly to be lethal. Accidental and malevolent exposure to high levels produces radiation burns and malignant cancers. At lower levels the incidence of cancers is reduced but not eliminated, and damage to the gonads is still a potential genetic hazard. In the face of a military strategy which is founded on nuclear weapons, and a civil strategy which requires nuclear power stations, it is prudent to be very much aware of the radioactive environment.

The basic particles involved are α-particles (helium nuclei), β-particles (fast electrons), γ-rays (photons of energy of the order of MeV), and neutrons. The most ionizing are α-particles, but they have as a consequence a very short range, being stopped by an aluminium plate a tenth of a millimetre or so thick. Such a plate would not stop β-particles, which are much less ionizing, but a few centimetres of aluminium would. More penetrating still are γ-rays, which require several centimetres of lead to absorb them. Most penetrating of all (because least ionizing) are neutrons, which entail shields of concrete metres thick. These particles are produced in the radioactive disintegration of certain atoms, and they appear at sea-level alongside other ionizing particles in cosmic rays. The ranges in water, which is similar to living tissue, of α, β, and γ-rays are given in Table 8.3.

Table 8.3 — Ranges of 1 MeV α, β and γ rays in Air and Water

Particle	Range in Air	Range in Water
α	5 mm	7 μm
β	0.4 m	4.3 mm
γ	100m	160 mm

The range of a particle in matter is determined primarily by its ability to ionize atoms, that is, to knock one or more electrons out of their orbits. Most of matter, even in the form of a solid, is empty space, and an uncharged nuclear particle like the neutron can count on travelling through tens of thousands of millions of atoms before sharing some of its energy in a collision. At the other extreme, the α-particle is something like a runaway bull-dozer. Being a helium nucleus it carries a charge of two elementary units, and every atom in its neighbourhood is subjected to the strong electric field of its two protons. To ionize an atom costs about 10eV. Add to that an amount of kinetic energy imparted to the ejected electron and the ionized atom, and the total cost of ionizing comes to about 30eV. A 3MeV α-particle therefore creates some 10^5 ions in its path before dissipating all its own kinetic energy. In a solid, a line of 10^5 atoms would be about 20 μm long, counting 2Å an atom. If each encounter results in ionization, the range of the α-particle would be 20 μm and that is not too far off what is observed. A β-particle carries only one elementary charge and being thousands of times lighter than the α-particle can more easily just bounce off a nucleus without losing energy. It is several hundred times less effective in ionizing, and consequently it has several hundred times greater range. Electromagnetic radiation such as x-rays and γ-rays, although electrically neutral, can interact with charged particles and can lose energy to them. In addition to the ability to knock electrons about the more energetic γ-rays can create, in the vicinity of atomic nuclei, electron-positron pairs but, all told, their passage through material is much easier than it is for charged particles, though less easy than for neutrons.

The ability to ionize is not the whole story as far as radiation damage is concerned. The production of ions is, after all, soon countered by recapture of the electrons. Though strong electric fields may be generated locally by the forcible separation of charge, they will be extremely transient affairs. What is more important is the disruption of molecules. Molecules are communities of nuclei held together by electron bonds and they can be disrupted either by the ejection of a nucleus by a massive projectile like the α-particle or neutron, or by the ejection of an electron in an ionizing event. The most important molecules are those which regulate and determine the activity and reproduction of the living cell, such as the nucleic acids DNA and RNA. Wrecking or mutating these molecules can kill the cell, wound it, or make it malfunction disastrously and produce cancers or genetic defects which can blight a new generation. Although all high-energy radiation produces such damage there are characteristic differences connected with concentration. Short-range radiation such as α-particles and β-particles produce local damage, and particularly sensitive organs, such as the reproductory ones, are at great risk should radioactive emitters of such particles get into the blood-stream and lodge there. X-rays and γ-rays produce non-local damage uniformly in a given tissue type, and neutrons similarly. Radioactive emitters of these rays affect the whole body, and in medical applications the whole volume of the body exposed to the beam is affected. Another difference is that α-particles and neutrons are much more effective at killing cells than other types of radiation, because they can eject atomic nuclei from molecules, which is a more disruptive effect than ionization.

The simplest measure of radioactivity is the number of disintegrations in unit time. The unit is the becquerel (Bq) equal to one disintegration per second. An older unit still in use is the curie (Ci), one curie being 3.7×10^{10} disintegrations per second. Such measures do not indicate the effect on a body which absorbs radiation, and yet this is what is of most significance.

Primarily, the effects on the absorber depend not on the number of particles absorbed per second but on the amount of energy absorbed by unit mass. The Standard International Unit for this is the gray, one gray being equal to one joule per kilogram ($1 J kg^{-1}$). An older unit very much in common use is the rad, equal to 100 ergs gm^{-1}. The conversion is 1 gray = 100 rads. We will work only with grays, in fact with micrograys per year ($\mu Gy\ y^{-1}$) or milligrays per year (m Gy y^{-1}). The natural background of radiation affecting man can be divided into three components.

(1) *Cosmic Radiation*

Although some variation with latitude and with sun-spot cycle (Sections 9.2 and 9.6) occurs, the variation on average is relatively small. At sea-level in the U.K., cosmic rays produce a radiation dose of about 280 μ Gy y^{-1}. This dose rate increases with altitude near sea-level by about 5 μ Gy y^{-1} every 100m. The neutron component of the sea-level dose rate is about 5 μ Gy y^{-1}.

(2) *External Terrestrial*

Rocks and soils contain uranium, thorium and potassium–40, all of which are gently radioactive. Since bricks and plaster, and other building materials are manufactured from such elements, usually those occurring locally, the external terrestrial background originates not only in the soils and underlying strata, but in the walls, ceilings and floors of our own houses. An average figure for the dose rate contributed by this component of the natural background is 380 μ Gy y^{-1} in the U.K., but this figure conceals substantial geographical variations. In granitic areas such as Cornwall and Aberdeen the background is some two or three times higher than low radioactive regions. In Aberdeen granite houses the average dose rate is 1,000 μ Gy y^{-1}, and in London brick houses it is only 400 μ Gy y^{-1}.

(3) *Internal Terrestrial*

Some radioactive isotopes such as potassium–40, carbon –14, polonium–210, rubidium–87, and daughter products of radon–222, affect the body from the inside by being incorporated into the blood, tissues and bone through the necessary acts of eating, drinking and breathing. One's body carries its own radioactivity around. Mostly this owes it origin to potassium–40, the radioactive isotope of potassium, which is a constituent of the blood. An average figure for the dose rate from internal sources is 210 μ Gy y^{-1}, but once again large variations can occur. One literally homely source of variation is the amount of ventilation in the living rooms of houses. One radioactive product of uranium in the brickwork is the rare gas radon–222 which can diffuse out into the air of the room. Radon is relatively harmless, but its daughter products are highly radioactive, and these can be attached to dust particles in the air, inhaled and deposited in the lungs. This effect constitutes one of the many hazards of hard-rock mining, and leads to an excess incidence of lung cancer. Though many times less hazardous in the home, the same processes occur naturally. Good ventilation, say two room changes of air an hour, reduces the radon-daughters effect by roughly a factor of ten compared with a low winter-induced ventilation rate of say 0.2 room changes an hour. Energy conservation schemes which aim to reduce the rate of air changes must balance the saving of energy against the possibly deleterious effect of the build-up of radon's daughters.

The total natural background is therefore 870 μ Gy y^{-1}, but with variations between 650 and 2,000 μ Gy y^{-1}. A round figure is 1 m Gy y^{-1}, that is one millijoule of radioactive-decay energy is absorbed by one kilogram every year. That figure represents an irreducible order of magnitude in our environment against which to set man-made radiation doses. Of the latter, the most important for the population as a whole derives from the medical use of radiation for diagnostic purposes and the treatment of disease. An estimate of the dose rate in the U.K. is 140 μ Gy y^{-1}, which is a considerable addition to the natural background. For comparison, the fall-out from atomic bombs in the forms of strontium-90 (in milk), caesium-137 (foods), tritium (water), carbon-14 (vegetation), amounts to 22 μ Gy y^{-1}. And by comparison with this, the effluent from nuclear power stations is insignificant. Effluent from nuclear reactors occurs in the form of long-lived isotopes, principally, the rare gas krypton-85, and tritium, which are global contaminants, and more local effluents such as tritium (again) and the rare gas argon-41. But all told, such contamination produced dose-rates of no more than 0.1 μ Gy y^{-1}, in 1974.

It is easy, with that figure of 0.1 μ Gy y^{-1} before one, to belittle the danger of radioactive effluents from nuclear reactor and reprocessing plants. That figure is a global one, calculated by assuming complete dispersal in the atmosphere and oceans. But complete dispersal takes time and since production is continuous such a concept is unrealistic. A gradient of radioactive pollution exists around all nuclear centres, and local effects are very significantly higher than global figures. Gaseous effluents are more rapidly dissipated than liquid ones, and it is with liquid discharges that we ought to be concerned. Radioactive atoms in liquid discharges can find their way into the local water supply but not significantly so. More important is their incorporation into marine plants, such as seaweed, and marine animals such as fish, oysters and other shellfish. Not only does this prevent the global dispersal taking place, it directly gets into the food chain of, if not the population at large, at least that section of the population with a particular taste for marine organisms of one sort or another. Such sections of the population are exposed to doses significantly higher than the reassuring global figure by factors from between a thousand and ten thousand! Cumberland seaweed from the Irish Sea, polluted by the Windscale reprocessing plant, is eaten in South Wales in the form of laverbread. Without dilution with other seaweed this results in a dose rate of 2 m Gy y^{-1}. This figure is twenty thousand times the global one. It is also twice the natural background and 40% of the dose limit of 5 m gray y^{-1} recommended by the International Commission on Radiological Protection in 1965. Fortunately for Welsh laverbread enthusiasts, the dilution of Cumberland seaweed with non-polluted seaweed results in a reduction of average dose by a factor of 4.2. Even more fortunately, the transport of Cumberland seaweed has become uneconomic recently because of increased rail-charges. Eating fish or being exposed to silt in an estuary or shallow sea can result in doses up to 10% of the I.C.R.P. limit. Thus the global figure may be meaningful to the average eskimo, but not to those who are unfortunate enough to live in the locality of a coastal reprocessing plant and who have an irresistible appetite for sea food. For them the background has effectively doubled.

Significantly, in 1977, the figure for radioactive waste had risen from 0.1 μ Gy y^{-1} to about 2.0 μ Gy y^{-1}, that is, by a factor of about 20. Most of this increase appears to be associated with radioactive caesium in the fish caught in the Irish Sea, North-West Scottish waters and the North Sea, whose origin is Windscale. This increase is disturbing. But it is important to put such radioactive pollution into perspective. The nuclear power industry is a product of high technology. It can monitor its pollution more accurately and more efficiently than old-fashioned industries which disgorge chemical pollutants into the atmosphere. It has to satisfy far higher standards than its competitors and it is continuously under public scrutiny. The unrestricted emission of sulphur and nitrogen oxides by the non-nuclear industry is far more deadly.

Much higher dose-rates than 1 m Gy y^{-1} can, of course, be encountered in certain occupations, but a safety limit is set for the average exposure of five times this value for the general public in the U.K. One naturally occurring hazard is found at jet-aircraft cruising altitude, where the cosmic ray background is about one hundred times higher than at sea-level, i.e 3 μ Gy h^{-1} or 26 m Gy y^{-1}. A trans-Atlantic flight of seven hours is equivalent to a year's exposure to fall-out! Table 8.4 summarizes the dose rates from various sources.

Table 8.4 — Radiation Dose-rates in the Environment

[Taken from G. A. M. Webb, 'Radiation Exposure of the Public — The Current Levels in the United Kingdom', *N.R.P.B.* — R24, (1974), with a rad-to-gray conversion.]

Source	Dose-Rate (m Gy y^{-1})		
Natural background			
1. Cosmic rays	0.28		
2. External terrestrial	0.38	0.87	
3. Internal terrestrial	0.21		
Medical irradiation		0.14	
Fall-out		0.022	(0.010)*
Miscellaneous (e.g. air travel)		0.003	(0.008)*
Occupational exposure		0.004	(0.007)*
Nuclear power industry		0.0001	(0.002)*
		TOTAL 1.04	

*Revised estimates from Taylor, F. E. and Webb, G. A. M. N.R.P.B. R77, London, H.M.S.O.

8.10 ENERGY FLOW IN THE ENVIRONMENT

Of prime concern to our industrial society, and of fundamental importance to all biological and to many physical processes, is the flow of energy in our environment. We live in a vast ocean of energy whose currents, waves and ripples sustain the countless busy activities of the planet with an immense, dispassionate and prodigal benificence.

The earth lives and feeds on energy from the sun, with a little help from its own radioactivity (a memory from a previous sun). In addition there is a contribution from its own gravitation and the gravitation of the moon and sun, and a tiny component of cosmic radiation from the rest of the universe. Of these, the flow of radiation from the sun is paramount. Over the lifetime of the earth it has induced life in the earth's surface and that life has laid down a record of the solar energy it once received hundreds of millions of years ago as chemical energy in coal, oil, gas and peat. At the present time that record is our civilization's main source of exploitable energy. These fossil fuels are the life savings of the earth. We have inherited a fortune and we are busy spending it as fast as we can. They are savings which cannot be replaced. As it is used, much of the fossil fuel energy, inevitably, is lost as heat. But some is converted into forms worth bequeathing: metals, plastics, chemicals, electronic crystals, buildings — semi-durable manifestations of our real wealth. Nevertheless, in time these riches will wear out and our fortune will have at last disappeared. Ultimately, fossil fuels and the wealth derived from them will be used up. That fact has an absolute quality about it which, as some of the few living beings produced by the sun that possess foresight, intelligence and imagination, we ought not to underrate nor subjugate to ephermal political expediency.

Tapping the virtually inexhaustible store of energy in the nuclei of the very heavy elements and in the very light elements has its own irresistable appeal. But even if the terrifying problems associated with the use of plutonium in nuclear reactors, and the disposal of nuclear wastes, are overcome in some way, and even if the technology of deuterium-tritium fusion is finally mastered, there still remains the problem of heat pollution mentioned in Section 8.6. In releasing nuclear energy we add to the energy flow. Unless we are prepared to tackle planetary engineering on the largest scale, the amount of heat which our generation and use of energy produces ought not to exceed about 1% of the input from the sun, that is it ought not to exceed $1.1 \times 10^9 \text{MW}$, which is some 100 times the present production. The solar input therefore imposes a limit on the artificial production of energy. The tapping of natural energy currents, on the other hand, does not impose limitations other than availability, and it is of some interest to appreciate the magnitudes of the various energy flows.

(1) *Solar Energy*

The rate of solar energy outside the atmosphere is 1.35kWm^{-2}. Taking the earth to be a circular disc of radius 6,360 km, allows us to calculate the area intercepting the sun's radiation. It is $1.27 \times 10^{14} m^2$. Consequently the total rate over the daylight hemisphere is $1.72 \times 10^{11} MW$. Of this $7.5 \times 10^{10} MW$ is absorbed at the surface, and $3.7 \times 10^{10} MW$ is absorbed in the atmosphere. These figures describe the basic energy input to the earth.

The radiation reaching the surface is mostly converted into heat, but a fraction is used by vegetable life on land and sea to photosynthesize carbohydrates out of carbon dioxide and water, and a fraction could be used by man to generate electricity in solar cells made out of semiconductors such as silicon or gallium arsenide. Covering the whole surface of the earth with solar cells operating at 10% efficiency would produce $7.5 \times 10^9 MW$, which is seven times the limit which heat pollution imposes on nuclear power. On the other hand, a more acceptable and realistic coverage of 0.1% (leaving room for all sorts of other things) leads to a power of $7.5 \times 10^6 MW$.

The power $7.5 \times 10^{10} MW$ is the average rate of surface absorption by the earth as a whole, and it corresponds to a continuous surface flux of $590 Wm^{-2}$. A given location on the earth's surface is not exposed to this flux continually because of the earth's rotation and because the incident energy of solar radiation is spread over a hemisphere and not a disk. Rotation causes the average flux to be halved since there is exposure to sunlight during, on average, only 12 hours of the day. Another factor of two is introduced by the curvature of the earth since the area of a hemisphere is twice that of a disc of the same radius. Thus the global average insolation at a point on the earth's surface is $150 W\ m^{-2}$ (roughly $300 W\ m^{-2}$ during the day and zero at night). In Europe the average is about $120 W\ m^{-2}$, in the U.S.A. it is $200 W\ m^{-2}$, and in the Sahara $260 W\ m^{-2}$.

(2) *Wind Power*

The troposphere obtains its energy mostly from condensation of water, evaporated from the oceans and land, and from the terrestrial beam. Energy contained in the latter is absorbed mostly by the first few metres of atmosphere, and convection spreads the energy through the troposphere. In Figure 8.5 about 137 units, comprised of 107 radiation and 30 evaporation and convection units, originating from the earth's surface, are absorbed in the atmosphere and mostly in the troposphere. Equating 100 units with the total incident solar power of $1.72 \times 10^{11} MW$ implies that $2.36 \times 10^{11} MW$ is absorbed by the troposphere, of which probably only some 4% or 5% is converted into horizontal atmospheric motion, say $1 \times 10^{10} MW$. Most of this will be in high altitude winds and inaccessible. If only the lowest 50m, i.e. a fraction 0.005 of the troposphere, is regarded as accessible, and kinetic energy is concentrated near the ground, enhancing the fraction of accessible energy to 0.05, then the potential power is $5 \times 10^8 MW$, reduced by 10% efficient windmills to $5 \times 10^7 MW$. But once more adopting a criterion of 0.1% surface coverage we arrive at a power of $5 \times 10^4 MW$ as being reasonably accessible. This is much lower than that available from solar cells. Nevertheless, in favourable locations wind power could be (and, of course, has been) an important source of energy locally.

Another way of estimating wind-power, is to assume a global average wind-speed of about 10km h^{-1}, say 3m s^{-1}. The kinetic energy per unit volume of air, density 1.29kg m^{-3}, is 5.8 Jm^{-3}, corresponding to an energy flux of 17.4W m^{-2} horizontal to the surface. The power produced by a 10% efficient windmill of sail-covered area 10^3m^2 would be a mere 1.7kW. Given a base for the windmill equal in area to a sail covered area, i.e 10^3m^2, and a total global area of $5.08 \times 10^{14}m^2$, the maximum number of windmills is 5.08×10^{11}. But to avoid down-wind screening of one windmill by another, only (say) 10% of this number is theoretically workable, i.e. 5.08×10^{10}. Again adopting an exploitation of 0.1% of theoretical maximum, we arrive at 5.08×10^7 windmills, corresponding to an average density of one windmill every 3 km over the globe, and a power output of 8.6×10^4MW. This is of the same order of magnitude as the previous estimate.

Other schemes for tapping wind-power involve indirect approaches. The most interesting of these perhaps is the plan to extract energy from the sea waves. A fraction (perhaps 10%) of the energy of winds is indeed converted into wave energy, and at favourable coastal sites it could provide a valuable source of power locally for a coastal community.

(3) *Photosynthesis*

The trapping of a fraction of the solar radiation by plants is literally vital for the whole of the biosphere. Of the 7.5×10^{10}MW absorbed by the earth's surface, perhaps about 1% is utilized in photosynthesis, by phytoplankton in the oceans and green vegetation on land. Most of what is utilized is dissipated by the organism in respiration and heat, but about a third is incorporated as growth, and available either as food for herbivores or as fuel. Thus a power of about 3×10^8MW is the food/fuel input *via* photosynthesis. Herbivores can utilize roughly a third of this plant energy, and about a tenth of the available energy is stored as growth. The latter is available to carnivores as food, but again only about a third can be utilized (Table 8.5). Herbivores are once removed from photosynthesis, carnivores twice removed. If all the plant life were edible by man a world population expending about 1×10^8MW could be supported. Since very roughly an individual's metabolism consumes energy at a rate of about 100W, the population would be 1×10^{12}. In the mid-seventies the world population was about 4×10^9, generating 4×10^5MW, or perhaps 4×10^4MW of useful muscle-power. A maximally populated world, on the same basis, would have a potential muscle-power of 1×10^7MW. On the other hand, if man insisted on being carnivorous the maximal population would drop to 1×10^{11} and the plant area required for a single individual with a taste only for meat would go up from $400m^2$ to $4000m^2$. A heavily populated world inevitably imposes a vegetarian diet as the norm.

Table 8.5 — Biological Energy Flow

Source	Energy Flux (Wm^{-2})	Percentage	Plant Area Requirement (m^2)
Solar radiation			
daylight average on plants	150	100	
Plants, absorption	75	50	
utilised in photosynthesis	2.5	1.7	
growth	0.75	0.51	
Herbivores, intake from plants	0.25	0.17	
growth	0.075	0.051	
Human requirements as herbivores			400
Carnivores, intake from herbivores	0.025	0.017	
growth	0.005	0.000034	
Human requirements as carnivores			4,000

Most energy could be extracted from vegetable matter by burning it. Simple burners are not difficult to make with an efficiency of 80%, so about 2×10^8MW of power from plant fuel is potentially available. If once more we limit realistic exploitation of 0.1% of the surface area the available power is 2×10^5MW, which is by no means insignificant.

(4) Geothermal energy

Radioactivity in the earth's crust produces heat and causes the temperature to rise with depth. Here and there on the earth's surface, that form of heat manifests itself in the form of hot springs or volcanic activity, and in such regions it has been used to provide useful energy. Some 1×10^3MW are presently produced in this way. Although this varies considerably from place to place, being concentrated near tectonic plate boundaries and younger strata, and, becoming weaker over the ancient continental cores, the geothermal heat current is ubiquitous. Recent measurements suggest a mean flux of energy of 50mW m^{-2}. Multiplying by the surface area of the globe, 5.08×10^{14} m^2, gives a global flow of about 2.5×10^7MW. Imposing once more a 0.1% surface coverage with a conversion efficiency of 10% suggests a potential of roughly 3×10^3MW. This puts geothermal energy on about the same footing as wind-power: not very significant globally, but worth exploiting at favourable sites.

(5) Waterfalls

Water is evaporated from the oceans and land and returned as rain. If we regard most of the water as coming from the oceans the whole process is approximately one of lifting water from sea-level to a height equal to the average elevation of the land, which is 825m, and consequently providing a source of gravitational energy. This energy is dissipated as kinetic energy of water flow and as heat during the time the rainwater finds its way back from the land to the sea, but some of this could be and, of course, is tapped in hydroelectric installations to provide usable energy. How much is available? The annual rainfall varies so much over the earth's surface that to consider an average value seems at first sight almost meaningless. In the Sahara it is practically zero. In the equatorial forests of South America, Africa and Indonesia it is above 2m. Nevertheless for our global estimate an average does have meaning, so for that purpose we will take the world average annual rainfall to be 0.3m. Coupling that with a total land area of $1.48 \times 10^{14} \mathrm{m}^2$ produces a mass of $4.4 \times 10^{16} \mathrm{kg}$ per year lifted to 825m, corresponding to a production of gravitational power of $1.1 \times 10^7 \mathrm{MW}$. Estimating that a third of this is convertible into electric power (allowing losses by evaporation, underground percolation, irrigation etc.) yields a potential supply of $3 \times 10^6 \mathrm{MW}$, provided that the building of dams and the flooding of valleys entailed by this is acceptable to society. The hydroelectric potential already exploited by the mid-seventies was about $2 \times 10^5 \mathrm{MW}$.

(6) *Tidal Power*

In Chapter 3 we saw that evidence existed which suggested that the earth's day had been lengthening by 1.6ms per century. If we take this figure to be a measure of the effect of tidal friction, the magnitude of the frictional power associated with the tides can easily be obtained from the knowledge of the rotational energy of the earth, $2.14 \times 10^{29} \mathrm{J}$. It turns out to be about $3 \times 10^6 \mathrm{MW}$. Some of this (about $5 \times 10^5 \mathrm{MW}$) is associated with earth tides, but it has been estimated by Jeffries and others that about $1 \times 10^6 \mathrm{MW}$ occur in the oceans and especially in shallow seas, bays and estuaries where the tidal range, commonly less than 1m in the oceans, reaches 10m and more. Of this about 6% is associated with coastal basins where exploitation is a realistic possibility. With a 10% conversion efficiency, the potential tidal power is about $6 \times 10^3 \mathrm{MW}$, which is globally on a par with geothermal power. Only about 1% of this potential has been tapped so far, mainly in France at La Rance Estuary. In fact, exploitation of tidal power is severely limited geographically. Only a dozen regions are of importance viz:

Europe: Irish Sea, English Channel, North Sea.
Asia: South China Sea, Yellow Sea, Sea of Okhotsk, Bering Sea,
            Malacca Strait.
Canada: Bay of Fundy, Hudson Strait, Hudson Bay, Fox Basin.

As in so many things, France has been the pioneer. The large capital investment at La Rance, tapping the tidal energy of the English Channel (more properly called in this circumstance La Manche) is paying off. The installation produces an annual average power of 62MW with a capacity of 240MW.

(7) *Earthquakes and Electric Currents*

A large earthquake releases an enormous amount of energy: one of magnitude 8.7 gives 5×10^{17}J. This energy happens to be equal to the average energy released by all earthquakes annually. Thus the average power for all types of earthquake is only 1.6×10^4MW, which is about a quarter that available from tides in coastal basins; but it is practically impossible to exploit.

A more frequent exhibition of terrestrial energetics than earthquakes are thunderstorms. An average lightning flash dissipates some 10^{10}J. If there is an average rate of one flash a minute in each of the 2,000 thunderstorms raging at any moment the total power is 3×10^5MW. It is not easy to see how this energy could be tapped efficiently. What could be tapped easily is the energy in the steady discharge current, but the power involved is very small, about 5×10^2MW.

(8) *Summary*

Table 8.6 summarizes the various energy flows in the environment.

Table 8.6 — Global Flow of Energy

Source	Power (MW)	%	Usable Power (MW)
Solar radiation, incident	1.72×10^{11}	—	—
reflected	6.0×10^{10}	—	—
absorbed in atmosphere	3.7×10^{10}	—	—
absorbed at surface	7.5×10^{10}	0.01 (a)	7.5×10^6
Atmospheric circulation	1×10^{10}	0.0005 (b)	5×10^4
Photosynthesis (Land and ocean)	3×10^8	0.08 (c)	2×10^5
Geothermal	2.5×10^7	0.01 (d)	3×10^3
Waterfall	1.1×10^7	33 (e)	3×10^6
Tidal, total	3.0×10^6	—	—
coastal basins	6×10^4	10 (f)	6×10^3
Atmospheric electricity, lightning	3×10^5	—	—
steady conduction	5×10^2		
Earthquakes	1.6×10^4	—	—
		TOTAL	1.1×10^7
Human metabolism (population 4×10^9)	4×10^5	10 (g)	4×10^4
World production (mid 70's, mostly fossil fuels)	2×10^7	33 (e)	7×10^6(h)

Notes on following page

Solar radiation is by far the most important. Large fractions of the sun's energy are absorbed at the earth's surface and in the atmosphere. Smaller fractions are used in photosynthesis, lifting water, and in atmospheric electricity. In addition there are geothermal, tidal and earthquake energies. And then there is the human metabolic rate, deriving from the sun's power *via* photosynthesis, and the present production of energy from fossil fuels. But at the earth's surface the dominant flow derives directly from the sun, and its magnitude of 7.5×10^{10} MW provides a basic measure to set against all other powers.

In estimating amounts of usable power, the author has adopted an arbitrary convention of limiting surface collection to 0.1% of the total global area. The reader will, no doubt, have his own ideas about estimating what fraction of a given source of energy is utilisable. The argument leading to 0.1% starts with an appreciation that only 29% of the earth's surface is land, of which only about a third is habitable. To avoid transportation of energy over large distances, with inevitable losses, energy must be produced as near as possible to where it is to be used, and that brings the area down to the inhabitable 10% of the total. To feed a population of 5×10^{10}, which is only twelve times larger than the present level, requires an area of about 5×10^{13} m^2 devoted to agriculture. But 5×10^{13} m^2 is 10% of the global surface already, therefore at best there is 1% of the global surface available on which there will be many other claims (towns, airfields, motorways, leisure areas, etc.). Thus, 0.1% devoted to energy production seems not too unreasonable. To those readers who are keen to get as much energy out of the system as possible, that figure will appear over-pessimistic. To others, who wonder what all the energy is needed for anyway, it will appear over-optimistic. In either case, we can conclude that the major sources of usable power are clearly direct solar energy and waterfall energy. Together, they appear to be capable of providing about 1×10^7 MW, which is more than our present requirements, now met by consuming fossil fuels. There is little doubt that if human society's need for energy amounts to orders of magnitude greater than presently produced, then the only way is *via* nuclear energy, preferably in the form of deuterium-tritium fusion. Now there appears to be no good reason for assuming other than that the energy requirement of human beings is infinite. If that is so, and controlled fusion becomes a possibility, it will soon become necessary to shift additional energy production off the planet on to the moon and Mars. An appetite for the infinite cannot be satisfied by a single planet, however bountiful.

Notes for Table 8.6 from preceeding page.
(a) Surface coverage of 0.1%, 10% efficiency.
(b) Surface density 0.1%, lower 50m of atmosphere, 5% of total, 10% efficiency.
(c) Surface area for fuel 0.1%, 80% efficiency.
(d) Surface coverage of 0.1%, 10% efficiency.
(e) Average efficiency of electricity production from rainfall.
(f) Conversion efficiency 10%.
(g) Sustained muscle-power capacity of 10%.
(h) Estimated from 1973 U.S.A. production of 2.4×10^6 MW multiplied by three.

At that stage, man will be forced to expand into space, or abandon his extravagant need of energy. As it is difficult to conceive of a cloud of dust remaining undispersed, so it is difficult to conceive of mankind forever bound to the earth. In the long run the need for energy will prove irresistable.

But in the short run there are formidable problems. All our technology is based on fossil-fuel energy. It is very easy to carry around oil and coal and burn it when we want energy. The energy comes already stored and ready for use. It is not so easy to carry around solar energy, waterfall energy, geothermal energy, wind energy and tidal energy. One gallon of gasoline or 4.4 litres of petrol is equivalent to 1.3×10^8 J of energy. A kilogram of coal is equivalent to 3.0×10^7 J. As against that, the carrying around of electrical energy needs special devices. The 2-volt lead-acid battery with an 80 ampere-hour capacity is equivalent to 5.8×10^5 J. Lighter and more robust batteries based on nickel and iron can be made but they produce smaller voltages (and to that extent smaller powers). It is not easy to match the energy per kilogram of the fossil fuels. To carry around energy from non-fossil fuel sources will require a new technology. It is highly probable that it will prove too expensive for individual users. Should that prove to be the case, we will lose some of the freedom that has come from the easy availability of energy. If technology is to have a value surely it must be that of conferring freedom of choice, of travel, of dwelling place, of clothing, of food; and if it is to have a chance we, as a human society, must give up nurturing expectations which are unrealistic.

A basic measure of energy expenditure is the power relative to that required by an individual human being to survive. A convenient measure of this is 100W. If the only energy available is that contained in food the energy level is unity. Taking the world production to be 2×10^7 MW and a population of 4×10^9 gives 5×10^3 W per capita, i.e. an energy level of 50. Now if the maximum production of energy is taken to be limited by heat pollution to about 1×10^9 MW, then the maximum energy level which the present population can attain is 2,500. But long before that production can be attained, the population will have increased by perhaps a factor of two at least, and the theoretical attainable energy level will drop to about 1,000 and more people still means even lower attainable energy level. In the U.S.A. however, the energy-level — an extrapolation of present day figures — will be 10,000. If the energy is to be equally available to all, the energy level of the U.S.A. has to drop by a factor of ten at least from what is at present that projected figure. Since energy is wasted in a fairly prodigal way in our industrial societies, a cut by a factor of ten will not be as painful as it sounds (Figure 8.8).

The measure of individual physical freedom (ethical freedom is quite another thing) in a society is therefore the energy level, defined as the production per capita divided by 100W. Production is limited by heat pollution. The only other factor is population. More people means less individual freedom. These are the basic physical facts and why energy has entered world politics. It will never be anything other than a primary issue ever again.

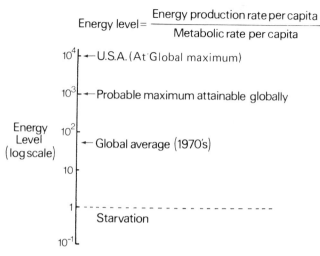

Figure 8.8 – Energy Level of Society.

SUGGESTED READING

Allen, C. W., 'Astrophysical Quantities', 3rd edition, (University of London, Athlone Press, 1973).

Cartwright, D. E., 'Oceanic Tides', (*Rep. Progr. in Phys.,* **40**, 665, 1977).

'*Energy and Power*', Scientific American, September 1971).

Few, A. A., 'Thunder', (p. 80, *Scientific American*, July 1975).

Gates, D. M., 'Atmospheric Electricity'' (*Rep. Progr. in Phys.,* **32**, 1, 1969).

Idso, S. B., 'Dust Storms', (p. 108, *Scientific American,* October, 1976).

Jeffreys, H., 'The Earth', 2nd ed., (Cambridge University Press, 1928).

Kauer, E. and Thalhammer, T., 'The potential of solar energy', (*Atomkernenergie*, Vol. 25, 1975).

Keeling, C. D., 'The Carbon-dioxide question', (p. 34, *Scientific American,* Jan. '78).

Lamb, H. H., 'Climate, Present, Past and Future', (Methuen, 1972).

Latham, J., 'Cloud Physics', (*Rep. Progr. In Phys.,* **32**, 1 1969).

Lomnitz, C. and L. 'Tangshan 1976', (*Nature,* **271**, 109, 1978).

Lynn, A. D., 'Air Pollution', (Addison-Wesley, 1976).

Murgatroyd, R. J., 'The Physics and Dynamics of the Stratosphere and Mesosphere', (*Rep. Progr. in Phys.,* **33**, 817, 1970).

Phillipson, J., 'Ecological Energetics', (Edward Arnold, 1966).

Resources and Man, 'U.S. Academy of Sciences', (Freeman, 1969).

Smith, P., 'The end of the expanding earth hypothesis' (*Nature,* **271**, 30).

Stow, C. D., 'Atmospheric Electricity' (*Rep. Progr. in Phys.,* **32**, 1 1969).

Sutton, O. G., 'Understanding Weather' (3rd edition, Penguin 1964).

9
The Celestial Sphere

The planets seem to interfere in their curves,
But nothing ever happens, no harm is done.
We may as well go patiently on with our life,
And look elsewhere than to stars and moon and sun
For the shocks and changes we need to keep us sane.
Robert Frost (1874–1963)
On Looking Up By Chance at the Constellations.

9.1 SPACE

If the atmosphere, thin though it is, protects us from the more vicious parts of the sun's radiation, it can do nothing about the impact which the night sky can make on our minds. If the atmosphere's window letting in visible light allows the development and well-being of life on the earth, that life has to be prepared to accept the impressive sight of the stars.

The vision is a unique part of our environment. The broad sash of the Milky Way, seen through powerful telescopes to be composed of countless individual stars; the cold twinkle of solitary stars forming the familiar patterns of the constellations; and the wanderers, brighter spots of light changing their position very slightly each night which we know as planets; all compose a physical presence felt even through the lamp-light of city streets. The effect of this image on magical, mystical and religious traditions of thought has been profound, and it cannot help but continue to exert a deep influence on our minds. To many, the hemisphere of stars has seemed to be a symbol of God. Our halting exploration of the moon, the closer planets Mercury, Venus and Mars, and the giants Jupiter and Saturn, is not going to change that. The solar system is, after all, a minute affair compared with the scale of the galaxy. It takes light 4.3 years to reach the nearest star, Alpha Centauri, 8.8 years to reach the brightest, Sirius, 587 years to reach the red giant Betelgeuse in Orion, and 30,000 years to reach the centre of our own galaxy. The symbol is not going to be tarnished with rusty familiarity through exploration of the stars, for that is plainly impossible.

But if we cannot reach the stars, the stars can reach us. Besides their strong effect on our imaginations, they impose their physical presence in several concrete ways, forever reminding us of our real connections with the immensity of the Milky Way galaxy and the universe beyond. This astral dimension in our physical environment goes beyond the twinkle of star-light (the twinkle incidentally, may be caused by processes in the eye as well as outside it) and, (since the sun is a star after all) beyond the glare of sun-light, or its pale reflection in the more gentle moonlight. It consists of influences for the most part invisible, and may even be the cause of something as ubiquitous as inertia.

Leaving aside the question of inertia's origin, and sticking to hard facts, we can scarcely avoid beginning our survey of the astral environment with the sun, whose proximity gives its idiosyncratic, as well as its ordinary behaviour, an importance to us on earth which is difficult to exaggerate. We have already discussed the solar radiation in previous chapters. A less familiar emanation is the solar wind.

9.2 THE SOLAR WIND

Since the beginning of this century it has been suspected by physicists that the solar system is a singularly draughty place, and this has been amply confirmed by experiments incorporated in space vehicles. On earth we have our magnetic field, as it were, to turn up our coat collar for us and protect us from the interplanetary blast, so we experience merely a few side effects, but that is all. If we were on the moon (and occasionally we are) there would be no protection of that sort, since the moon does not have a magnetic field. The importance of a magnetic field is that it does not allow charged particles to travel in straight lines, and our 'coat-collar' works because the wind is a plasma wind, that is, a wind of charged particles (Figure 9.1).

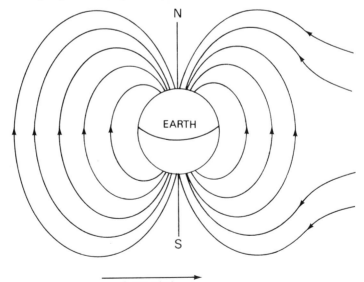

Figure 9.1 – Distortion of magnetosphere.

The nature of this wind is a rapid stream of ionized hydrogen emanating from the sun continuously. The electrons and protons speed past the earth at speeds of 400kms^{-1}, and eventually lose themselves in interstellar space. Those that hit the earth's field distort the magnetosphere, replenish the Van Allen radiation belts, and on occasion, cause auroral displays by pushing electrons down and ionizing the atoms of our upper atmosphere. On average about 10^{12} protons and electrons pass through a square metre per second. It is just as well that this flux of particles is so weak. Plasmas are intrinsically very good conductors of heat, and an intense particle concentration would allow heat from the sun to reach us *via* conduction as well as radiation. In effect we would then be part of the corona, the thin outer atmosphere of the sun – which would not be comfortable.

That the wind has a fairly steady component is shown dramatically by the striking appearance of comets. Comets frighten everybody, even theoretical physicists, and dire stories are told about them. Edmund Halley was the first to apply Newton's theory of gravitation to a comet, now called Halley's comet, in 1682, and predicted it should have an elliptical orbit and to return every 75 years. Records confirm this. The first well-authenticated observation, by Chinese astronomers, was 239 BC. and the most recent was 1910, and we confidently expect it to be back in 1986. Not all comets are so well-behaved: some appear for a few revolutions and then disappear, nobody knows why. Most comets reappear more frequently than Halley's, the average period being about 7 years. An increasing number are being discovered, at a rate of about 3 a year, which have periods in excess of 100 years (Figure 9.2).

There are, on average, two comets visible annually. A comet consists of a head which exceeds 20,000km across, and which contains within it a central condensation 2,000km wide, which itself has a nucleus of diameter 10 km. The nucleus is believed to be solid, and composed of frozen gases and dust particles, which evaporate when the comet comes within about 1.7 A.U. (1 A.U. or astronomical unit equals the mean sun-earth distance, 1.5×10^8km). The result is a tail which stretches over 10^7km directly away from the sun. Indeed, there are often two tails, which have v-shape formation.

One tail glows as a result of the emission of light by ionized gas molecules (mainly ionized carbon monoxide, CO^+), the other appears to be yellow because it is illuminated by reflected sunlight glowing on dust particles of about one micron (10^{-6}m) in diameter.

The origin of comets is obscure but, whatever their source, they act as wind-socks for the solar system. The ionized gas in the comet interacts strongly with the solar wind and gets pushed away from the sun. The dust, being un-ionized, does not get pushed as hard by the solar wind, but this is compensated for by the stream of photons which, in reflecting from and being absorbed by the dust particles, exerts a radiation pressure. Thus two tails can develop, one a tail of ionized gas, the other a tail of dust particles. At one time it was believed that radiation pressure on its own could account for comet tails, but this is not so. It is now thought that comet tails, particularly the ionized tails, are striking consequences of the solar wind.

That there occur gusts of the solar wind in addition to the steady breeze is shown by the simultaneous reduction in the intensity of cosmic rays. The latter consist of charged particles and so are affected by the magnetic field captured by the solar wind (Figure 9.3). The bombardment of the earth by cosmic rays can be halved at the peak of sun-spot activity. Gusts of solar wind also produce magnetic storms, in which the earth's magnetic field is perturbed, and earthly communication systems − radio, telephone, etc. − are disrupted. The frequency of magnetic-disturbed days per year ranges between roughly 40 and 80 in an eleven-year cycle corresponding to the sun-spot cycle. Severe storms occur on two or three days a year at the height of solar activity. A disturbance on earth can be associated with a flare on the surface of the sun because it follows the flare by a day or two, which is just the time taken for particles to travel between sun and earth.

How the solar wind originates is not fully understood. Clearly it is affected by solar activity on the surface of the sun. Activity manifests itself visually as

Figure 9.2 (*below*) − Halley's comet.
(Photo. Slipher 1910. Crown Copyright, Science Museum, London)

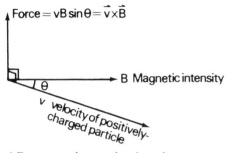

Force $= vB\sin\theta = \vec{v} \times \vec{B}$

B Magnetic intensity

v velocity of positively-charged particle

a) Force exerted on moving charge by magnetic field.

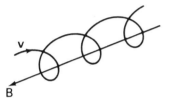

b) Corkscrew motion in field. Particle 'captured' by line of force.

c) Lines of magnetic force captured by outward flow of charged particles. (Converse of (b)).

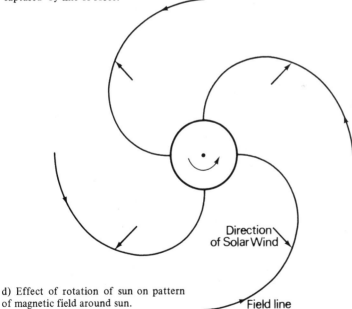

d) Effect of rotation of sun on pattern of magnetic field around sun.

Direction of Solar Wind

Field line

Figure 9.3 – Interaction between magnetic field and moving charged particles.

sunspots, flares, filaments and prominences (Figure 9.4). Sunspots are relatively cool patches of the sun's surface, rotating with the sun, but often initiating immensely hot flares above them, which emit bursts of intense light, x-rays and ionized hydrogen. The bursts of light and x-rays reach the earth in a few minutes, the burst of particles a day or so later. The number of sunspots varies between zero and about 100 at maximum activity. The interval between bouts of maximum activity varies between 9 and 13 years roughly, with a mean of 11.04 years. It is possible that solar activity may influence the weather, besides inducing magnetic storms and deflecting cosmic rays.

Somehow energy, produced deep inside the sun by hydrogen fusion and transmitted to the surface by radiation, and near the surface by convection, excites the corona to an immensely high temperature (about 10^6K), perhaps through acoustic vibrations, in such a way as to produce a solar wind. The solar wind provides a material (as distinct from radiative) connection between conditions on the surface of the sun and the earth. Thus is our capacity to communicate around the globe delicately dependent upon the humour of our immense companion of space. Even at a distance of 1.5×10^8km the earth is by no means insulated from the vast, explosive events which occur on the turbulent surface of the sun.

9.3 SOLAR TIDES

As the sun and moon raise the tides on the earth, so does the earth raise tides on the sun — plasma tides rather than oceanic tides, of course. Indeed, each planet exerts its gravitational tension on the sun and each planet raises its own tide of solar material. The law of gravitational torques and tensions (Section 3.6) relates the strength to the mass divided by the cube of the distance, and this quantity is shown in Table 9.1 for each planet.

Table 9.1 — Strengths of Gravitational Torques and Tensions at the Sun of the Planets relative to the Earth

Planet	Mass	Average dist to Sun (A.U)	M/R^3
Mercury	0.0532	0.39	0.897
Venus	0.817	0.72	2.19
Earth + Moon	1.00 + 0.012	1.00	1.012
Mars	0.107	1.52	0.030
Jupiter	318.	5.20	2.26
Saturn	95.1	9.54	0.11
Uranus	14.5	19.2	0.0020
Neptune	17.2	30.1	0.00063
Pluto	0.18	39.5	0.0000029

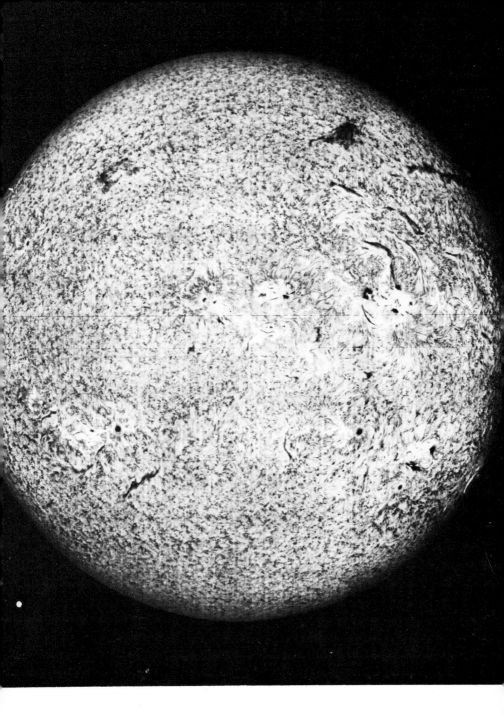

Figure 9.4(a) (*above*) − The sun photographed in hydrogen light.
(Mt. Wilson, Science Museum, London)
Figure 9.4(b) (*following page*) Large eruptive prominence on sun.
(Evershed, 1916, Science Museum, London)

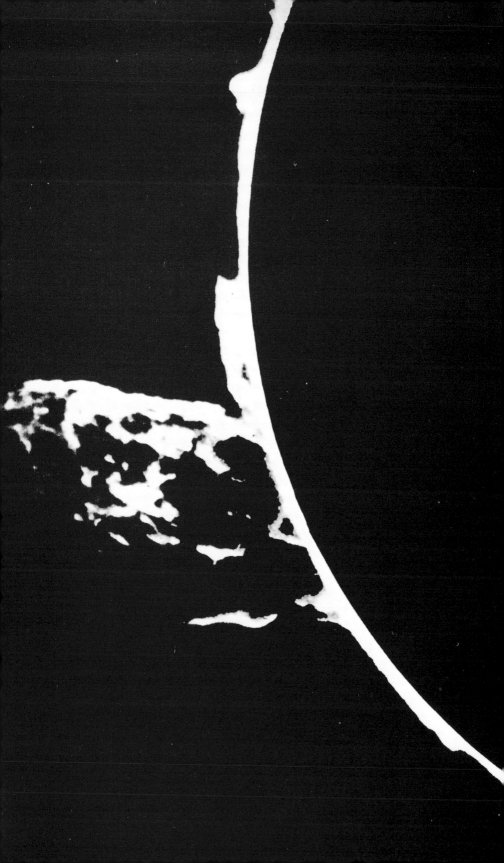

The interesting fact emerges that only five planets are effective in raising solar tides. These are Mercury, Venus, Earth, Jupiter and Saturn. Mercury, though small, is very close to the sun, and exerts about the same influence as does the Earth. Venus, because it is closer to the sun, and Jupiter, because of its immense mass, each exert an influence twice that of the Earth. The effect of Saturn is relatively small, and the pulls of Mars and the planets beyond Saturn range from small to insignificant.

Just as there are spring tides on Earth twice a month when the sun and moon are aligned, so there are spring tides periodically on the sun whenever the five tide-raising planets are aligned. This period was discovered by K. D. Wood in 1972 to be 11.08 years (discounting the comparatively rapid rotation of Mercury). Every eleven years Venus, Earth, Jupiter and Saturn line up and raise spring tides in the sun. The sun has a tidal month of 22 years! The eleven year cycle, Wood points out, corresponds with the sun-spot cycle. Moreover, the years of peak solar tides correspond with years of peak sunspot number.

If this remarkable coincidence of period reflects a casual connection between solar tides and solar activity, which is very plausible, then the oft-ridiculed planetary influences of astrology have to be taken more seriously. Although a direct effect on the terrestrial environment of the planets is difficult to conceive, the sensitivity of that environment to the state of the sun's surface opens up a causal channel through which the basically weak planetary forces may become amplified and affect the earth indirectly. If sunspots affect radio communication on earth, the rate of incidence of cosmic rays, and even the weather; and if sunspots are somehow produced by planetary tides; then is our environment truly governed by planetary configurations. It has been suggested by J. Gribbon and S. Plagemann (1976) that the solar channel transmits a trigger for earthquakes:

planetary alignment → solar tides → sunspot maxima → solar wind → atmospheric disturbance → earthquakes.

They warn us to watch out for the next planetary alignment due in 1982.

Solar tides are not the only manifestations of planetary configurations. The focal point of the solar system is not the sun: it is the centre-of-mass. The position of the planets affects the position of the sun with respect to this point, and this in turn affects the strength of the centrifugal and Coriolis' forces which operate on the matter composing the sun while it orbits about the centre-of-mass. How this effect, or the production of solar tides generates sunspots is not known, but these interactions between the sun and the rest of the solar system are of enormous significance to the earth and its environment.

9.4 INTERPLANETARY DEBRIS

The decimation of grouse on the Scottish moors, to the delectation of the hunter and the gourmet in us all, begins on 'the glorious twelfth' of August, and is appropriately celebrated about that time by a spectacular display of shooting stars. It is conceivable that grouse may have grouse-law as we have folk-law, and if so, the likelihood is high that the August 12 display of meteors enters into their imagery with powerful awesomeness. In us the familiarity, and the make-a-wish prettiness of shooting stars, has long since annulled any similar feelings. And yet, perhaps the grouses's conjectured response is more rational, more in tune with the nature of things, than our own.

Every shooting star we see is the manifestation of a collision of the earth with an extra-terrestrial object. We call such streaks of light meteors, and we know they are caused by grains of rock called meteoroids which travel so quickly through the atmosphere they heat themselves, and the air, to white heat. Those meteoroids which are too tiny to produce such heat and light are called micrometeorites, and they drift down to the surface of the earth as dust. Those meteoroids which are large enough to avoid being burnt up, reach the earth's surface as meteorites, and turn out to be chunks of stone or iron weighing typically several kilograms. That such things drop out of the sky is rather daunting, but even worse is the palpable evidence of dozens of huge craters, some many kilometres in diameter, scattered about the earth's surface, recording, like their kin on the moon, the collossal impact of giant bodies up to tens and hundreds of thousands of kilograms. Such bodies may indeed have been asteroids or small comets.

The occurrence of gigantic collisions may not be as rare as one would like to think. There were two events in Siberia this century — one the devastation of a forest in 1908, the other in 1947 in which about 100 craters were produced by a mid-air explosion of a meteoroid. On the other hand the Barringer meteorite crater in Arizona is a comfortable 20,000 years old. But reports of disasters involving sizeable meteorites occur with disturbing frequency throughout recorded history.

The less threatening collisions go on continually. On an average night there will be about 10 meteors an hour, but on those special nights which occur at various times throughout the year the rate may go up to 40 an hour or more. These meteor showers occur when the earth's orbit intersects a meteoroid stream, which itself orbits the sun. Such is the Perseid stream, which may or may not influence grouse-law by its spectacular manifestations for a week on either side of August 12. Many streams are clearly associated with the orbits of comets, but others, the Quadrantid stream (January 2-4) and the Geminid stream (December 8-15), have not so far been associated with a comet. Whether the meteoric material is derived from showers or from sporadic collisions, the total amount picked up by the earth each day is an impressive 10 tons. But even this is small in comparison with an estimated 400 tons of micrometeorite material a day. Interplanetary space is extraordinarily dusty. Indeed, the dust which inhabits space between the sun and the earth's orbit is present in sufficient quantity to scatter a detectable amount of sunlight. This scattered light is dimly visible at dusk and dawn and is called zodiacal light. Dust lying near the earth is swept up by the earth's gravitational field. The earth, it seems, is a gigantic, interplanetary vacuum cleaner.

It is a curious fact that more meteors occur after midnight than before. The reason why this asymmetry is observed, is bound up with the earth's motion around the sun. From the viewpoint of the Pole Star the earth orbits anticlockwise and also rotates about its axis anticlockwise. Thus, the hemisphere which points to the front (however bluntly) is always the hemisphere experiencing times between midnight and noon. Being at the front, it takes the brunt of any collisions that are going; and so meteors are more likely to occur between midnight and noon, though visible only between midnight and dawn.

The constitution of meteorites is of considerable, if melancholy, interest because it appears to be very much what one would expect of earth fragments, were our planet to blow up. Of the meteorites large enough to be seen falling and found (about two per day over the whole earth), the vast majority are stoney and about 10% are stoney-iron or iron. The evidence we have concerning the earth's interior suggests that its mantle is stone and its core is made of iron. It is possible therefore that meteorites are the remnants of a planet. Besides carrying the memory of the origin of the solar system through their radioactivity (Chapter 5), perhaps they also bear mute witnesses to an archaic, celestial disaster.

9.5 COLLISIONS

Though we cannot rule out the chance of the earth suffering a really serious collision with a sizeable member of the solar system, we can certainly say that it would be a highly unlikely event. There is, of course, no chance whatsoever of hitting any of the other major planets, since their orbits do not intersect the earth's. Our nearest neighbour, Venus never comes closer than about 40 million kilometres; and our superior neighbour Mars (superior because further out from the sun) is never closer to us than about 80 million kilometres. It is almost certain that the largest bodies of the solar system have, over the thousands of millions of years since their origin, agreed dynamically amongst themselves which orbit each should follow in order that perturbations be as small as possible. The same cannot be said of the asteroids, the name we give to the minor planets.

9.5.1 The asteroids

In 1801 Ceres, the largest of the asteroids, was discovered by Piazzi, and in 1802 Pallas and Vesta, the two next largest, were discovered by Olbers. This astronomer, famous for his paradox, speculated that these asteroids were fragments of a former planet whose orbit lay between the orbits of Mars and Jupiter, and his speculation stands today. The origin of the asteroids remains a mystery.

In 1972 the list of asteroids with determined orbits numbered 1779. Most of them are chunks of rock about 10km in radius, but the range of size stretches from radii 380km for Ceres and 240km for Pallas and Vesta, down to less than a kilometre. Their orbits are more or less in the plane of the ecliptic and are on the whole gently elliptical. Ceres and Vesta have nearly circular orbits, but Pallas is something of a maverick in having a pronouncedly elliptical orbit inclined to the plane of the ecliptic by as much as $35°$. On the whole the behaviour of the bigger asteroids fits into the general pattern of the solar system and they represent very little hazard to navigation. It is the behaviour of some of the smaller asteroids that constitutes an interesting argument in favour of supporting space technology.

The orbit of Eros is not more elliptical than that of Pallas; but it is much nearer the earth, nearer on average than even Mars, and so its elliptical path brings it every seven years or so to a point a mere 25 million kilometres away. This is comfortably distant from what is little more than an outsize boulder − 7km × 19km × 30km. Far less comfortable were the near misses of three small asteroids, each a few hundred metres in radius, irregularly shaped and tumbling over and over as they hurtled past the earth. In 1932 Apollo missed us by some 3 million kilometres; in 1936 Adonis came within 2 million kilometres; and in 1937 Hermes achieved a record, which still stands in 1977, of about 750,000km − about twice the distance to the moon. Apollo, Adonis and a dozen or so companions regularly repeat this alarming behaviour. Another potential menace whose orbit is extremely well plotted is Icarus, a body about 700 metres in radius. Its average distance from the sun is the same as the earth's and on that score Icarus is a very near neighbour indeed. It has, however, a highly elliptical orbit which takes it within the orbit of Mercury and beyond the orbit of Mars, and as the name suggests, it has the distinction of getting closer to the sun than any other asteroid. Fortunately, (since, unlike its classical namesake it survives this solar proximity, it ought perhaps to be called Daedalus) the plane of its orbit is inclined to the plane of the earth's orbit by the large angle of $23°$; so for the most part it is well out of the earth's way. In 1949, when it was discovered, it missed the earth by some 6 million kilometres. The orbits do not quite intersect but it would take little in the way of perturbation to make them do so. Icarus goes around the sun in 408 days, or 1.12 years. That means that every 19 years Icarus and the earth should be in close proximity. This occurred in 1968 and should recur in 1987 (one year after Halley's comet). It would be an interesting challenge to space engineering if astronomers were to discover that Icarus were heading for a collision with the earth. They would have to contemplate a landing on Icarus, mounting a powerful rocket engine, and blasting it off course. Fortunately, in 1987 it will miss the earth by a comfortable 6×10^6km.

Relative to the size of the solar system the earth is tiny, and an asteroid is far tinier. Even though most of the matter in the solar system orbits very roughly in the same plane, there is still a great deal of space, and the chance of a serious collision is small. We must, however, bear in mind that the probability of a serious collision depends not only on celestial whimsicalities, but also on what we mean by serious. Meteorites are a daily occurrence. The vast majority are far too small to detect before they hit the atmosphere. Even the big ones which have produced the terrestrial craters must have been only a few tens of metres in diameter, much smaller than those asteroids observable with powerful telescopes. Yet such objects potentially can produce immense devastation. But the chance of a crater-type collision is many times smaller than a major earthquake, and we can do nothing about either of these hazards at the present time. Earthquakes are far more frequent than damaging meteorites, and yet the effect of even the most powerful is limited to a relatively small area of the globe, whereas a collision with an asteroid would be devastating over a much greater range. The chance of such a disaster happening, or even threatening, may be very small but, the terrible magnitude of destruction which would be involved, makes the constant monitoring of the orbits of the larger pieces of debris inhabiting space — asteroids and comets — a prudent, as well as scientifically interesting occupation.

9.5.2 Stellar collisions

Besides travelling around the sun at a speed of about 30kms^{-1}, the earth is dragged along through interstellar space at the sun's speed relative to local stars at about 10kms^{-1}. What is the chance of the sun colliding with something? The answer is extremely reassuring. Interstellar space is very empty, so the main candidate for a collision is another star. In the neighbourhood of the sun there is approximately one star in ten cubic parsecs, corresponding to a distance between stars of about 10 light-years. If it takes light, travelling at $3 \times 10^5 \text{kms}^{-1}$, 10 years to get from star to star, it will take the sun ten thousand times longer. Thus the shortest conceivable time to a collision is about 10^5 years. But even then, the likelihood is high that the sun will miss the star it is heading towards by a distance of the order of a light-year, far too distant to produce a noticeable effect. Chandrasekhar, the astrophysicist, has estimated that it will take 10^{10} years for the sun to be deflected $1°$ as a result of a stellar collision. Since this time is greater than the age of the sun, which is somewhat greater than the age of meteorites (4.6×10^9 years), we may safely leave the topic of astral collisions to the astrophysicists.

Comets, asteroids and meteoroids remain the biggest hazard. Unfortunately, nobody knows how often really big collisions occur — one a century? — ten a century? Saying that earthquakes are more frequent, so forget about things hurtling down from outer space, is scarcely reassuring. Two words of comfort are perhaps in order; one is that the chances are of the missile hitting ocean rather than land; the other is that the atmosphere affords a certain amount of protection. On April 10th 1972, a meteoroid weighing an estimated 1,000 tons bounced spectacularly off the top of the atmosphere, glowing 60km above the state of Montana: had it been a few kilometres lower, with a slightly different angle of approach, the neighbouring state of Alberta might have acquired, in dramatic circumstances, an unwanted crater. That time, the atmosphere, though extremely tenuous at 60km, was shield enough and protected us. How many times that happens is known perhaps only to the military air forces of the world, looking out for one another's missiles.

9.6 COSMIC RADIATION

The earth is bathed in radiation from the sun, the galaxy, and the rest of the universe, out to its furthest reaches. Visible and invisible messengers, photons and particles, taking minutes, years or eons over their journey, bear physical witness to both commonplace and great events occurring throughout space and time, and connect the earth through their nature, their energy and momentum, to the cosmic societies of stars. Much of their import is of powers too vast in human terms to be safely borne without the beneficial mediation of the atmosphere, whose separation of the life-giving radiations from the death-giving phenomena is vital to us. But if the atmosphere succeeds in blocking our delicate ears to the more strident of the *harmoni mundi* (the harmonies of the world), its success rests solely on the diluting effect of the wholesale expansion of the universe, without which, as Olbers' paradox warns, the whole of the celestial sphere would be as bright as a star, and no amount of protection by a 100km thick skin of air would save us.

That 100km of air is desperately thin. People commute that distance twice a day! It has to cope with a bombardment of quantum particles whose energy ranges from 10^{-9}eV, a radio photon of frequency 1 MHz, to 10^{20}eV, the most energetic cosmic ray primary particle; and it copes so well, that we have to go to enormous expense to send artificial satellites, or moon rockets in order to rise above the atmosphere and take a look, because only a few spectral windows let in electromagnetic radiation; one in the visible, of course; seven narrow windows in the infrared between 1 μm and 20μm and everything beyond 800 μm.

Everything beyond 800 μm means the part of the electromagnetic spectrum we use in radio and radar. In terms of frequency it means everything with frequency less than 400 GHz (400 thousand million waves per second). In the radio band 1 MHz to 500 MHz we are exposed continuously to a highly complex radiation field whose psychological and physical influence on man is difficult to assess, but whose origin is far from being heavenly. It consists of shortwave and F.M. radio and V.H.F. television broadcasts. All of this, plus a good deal of radio noise emanating from electric drills, domestic machinery, railways and a thousand other things, plus a contribution from the ionosphere, practically drown the weak radio signals from space, but nevertheless the cosmic radio background is there. It speaks of electrons spiralling around magnetic fields, in the sun, in radio stars, in radio galaxies, but the ear has to be finely tuned to hear it. It underlies all our favourite programmes.

Nowhere is the intensity of a man-made terrestrial radiation field more evident than in the vicinity of a powerful radar transmitter, emitting in the microwave band, 500 MHz − 500 GHz. Living electrical conductors, like birds, sheep and people, may be invisibly cooked to a turn, as in a microwave oven, should they get too close. The thousand-foot antenna of the Arecibo Observatory in Puerto Rico, in terms of black-body radiation, would look to an extra-terrestrial observer like a hot spot on the earth at a temperature of some 10^{23}K. Most places are reasonably well out of the way of radar beams, and their terrestrial microwaves come from the atmosphere, where water and oxygen molecules radiate busily. But once again, behind all this earthly activity, lies a celestial murmur − the cosmic microwave background. Between about 1 GHz and 100 GHz the universe is relatively quiet, and there is only the all-pervading background of a radiation which, we believe, reaches us from the beginning of space and time itself. On the earth's surface, because of atmospheric noise, we detect it most easily between 1 GHz and 10 GHz. The way its intensity varies with frequency shows that it is pure blackbody radiation corresponding to a temperature of 2.76K. The temperature, plus the observation that its intensity is remarkably independent upon direction in space, are consistent with the theory that the microwave background radiation was first produced in the big bang which started the present phase of the universe off some 10^{10} years ago. Universal expansion has reduced its temperature to near absolute zero, besides making our nights dark. The ear-splitting roar of creation has become an all-pervading whisper (Figure 9.5).

Naturally, in the infrared and visible regions the sun tends to dominate everything. Outside the solar system, the brightest infrared objects are large clouds of dust in the galaxy, and also the galactic centre itself. The latter is optically invisible because of intervening dust, but infrared radiation at wavelengths between 20 μm and 100 μm get through conspicuously, as do radio waves. The crowded centre of our galaxy communicates with us only through the medium of the longer electromagnetic waves. The galaxy also contains many sources of ultraviolet light, x-rays, and γ-rays up to a photon energy of 100MeV, which join with the sun's emission to keep the atom at the top of our atmosphere effectively ionized. Fortunately for us, very little of this lethal radiation reaches the ground.

At all wavelengths, we find that there is a steady background intensity which derives from the sun, the galaxy, and beyond the galaxy. But superimposed on this are sporadic bursts of radiation which swamp the background completely. Bursts of radio and microwaves emanate from sunspots, and powerful bursts of x-rays shoot out from solar flares. Recently it has been discovered that bursts of γ-rays hit the solar system from outside. Lasting from 0.1 seconds to 100 seconds, they produce a count rate reaching levels of 1,000 photons per second, which has to be compared with an average of one or two photons per second. One of these bursts were recorded by the manned spacecraft Apollo 16 on its return from the moon on April 27th, 1972. They happen at a rate of about one a month. Their origin is uncertain, possibly extragalactic, probably galactic. γ-rays destroy life. Such bursts are dangerous to living entities outside the protective screen of our atmosphere.

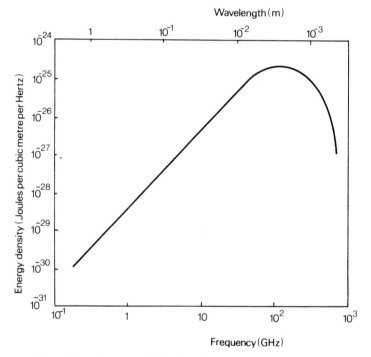

Figure 9.5 — Spectrum of Cosmic Microwave Bacground. It agrees well with the spectrum of a blackbody at 2.76k.

The most spectacular sources of sporadic bursts of radiation reaching the earth are undoubtedly the supernova of our galaxy, because they are visible. A nova is a star which suddenly becomes brighter and may become the brightest star in the sky, a condition which lasts for weeks rather than months, after which it returns to comparative obscurity. This is a fairly common occurrence in the evolution of a star and during its outburst the star is believed to throw off a small fraction of its mass into the surrounding interstellar space or towards a companion star. Six particularly bright nova are recorded this century, the most recent one being in 1975. The supernova is on a different scale altogether. At its peak of brilliance it can outshine a galaxy, and can persist for months or even years. It is believed to be an exploding star. It is also thought to be the origin of high-energy cosmic rays, and element heavier than iron, besides producing radio and x-ray sources, and runaway stars which hurtle through the galaxy. Unlike a nova, the supernova ends up as a neutron star, or even a black hole. Less than one in a century have been recorded. The most recent in the year 1604, was recorded by Kepler (1571-1630) and also by Chinese and Korean astronomers. Bright stars, thought to be supernovae, were recorded in 1572, 1181, 1054, 1006, 393 and 185. The 1572 star was particularly significant for European astronomy, in that it stimulated the young Danish alchemist Tycho Brahe (1546-1601) to make a careful observation of this extraordinary object, which was a blasphemous example of change in the perfect, unchanging heavens of Aristotelian cosmology. So impressed was the King of Denmark with Tycho's subsequent account that he gave Tycho the island of Hven, where, Danish nobleman that he was, Tycho built the extraordinary Castle of Uraniborg, and created a thriving school of observational astronomy (Figure 9.6). The 1572 supernova induced the birth of modern astronomy. Once as bright as Venus, it is now a nondescript radio source a few thousand light-years away on the edge of our galaxy in the direction of the constellation Cassiopeia (the W-formation high overhead in the northern winter). Exactly 400 years after its inspiring image, man was on the moon, in his first phase of space exploration.

9.7 COSMIC RAYS

Besides being bathed in electromagnetic radiation, the earth is scoured with cosmic rays, fast-moving particles from the sun and from the galaxy. Along with the earth's own radioactivity, cosmic rays are responsible for the background tick of the Geiger counter, the ionization of the air at ground level, and the steady bombardment of the molecular structures of living things. Again, the atmosphere acts as a protective shield, cutting out the numerous low-energy particles. Because cosmic ray particles are charged, the earth's magnetic field can also help by deflecting the slower ones (Figure 9.7). This works at the galactic level too, because the galaxy is permeated with a weak magnetic field some hundred thousand times smaller than the earth's. Cosmic rays originating within the galaxy take about a million years to escape; even though most of them travel at speeds close to the speed of light, and would fly out of the galaxy in about a tenth or a hundredth of that time if they were allowed to travel in straight lines.

The cosmic rays which hit the top of our atmosphere, the so-called primary particles to distinguish them from the secondary particles produced in a nuclear collision, are basically fast electrons and the nuclei of the elements. Their composition roughly matches the natural abundance of the elements, as found in the rest of the universe, but not quite the same − hydrogen is less abundant and the light elements lithium, beryllium and boron are very much more abundant. For every 1000 protons (hydrogen) there are 160 α-particles (helium

EFFIGIES TYCHONIS BRAHE O. F.
ÆDIFICII ET INSTRUMENTORVM
ASTRONOMICORVM STRVCTORIS
Aº DOMINI 1587 ÆTATIS SVÆ 40

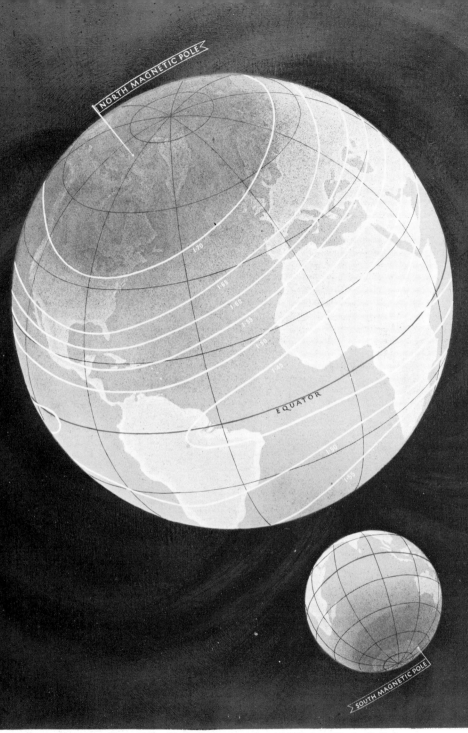

Preceeding page: Figure 9.6 – Tycho's mural quadrant 1587, from Tycho Brahe's
Astronomia Instauratae Mechanicae, 1602.

(Science Museum, London)

Above: Figure 9.7 – Cosmic ray distribution over globe. Figures on curves are
number of ion pairs cm^{-3} s^{-1}.

(Science Museum, London)

nuclei), one or two of the lighter elements such as carbon, nitrogen and oxygen, and only one in a thousand chance of there being say a copper nuclei. If we are right in thinking that they originate in supernova, they represent the distribution to the rest of the galaxy of the chemical output of vast stellar factories, some of which is even exported to other galaxies.

We have already noted that the sunspot cycle, through the solar wind, exerts a sort of periodic 'import control' by deflecting cosmic rays by means of the magnetic field trapped inside the relatively slow-moving stream of ionized hydrogen which forms the solar wind. At sunspot minimum there are about 6,000 cosmic ray particles per square metre per second flying through the solar system; at sunspot maximum that flux is halved. The average energy of one particle is a substantial 10 GeV, but energies are observed to be as high as 10^{20} eV. To get energies like 10 GeV we have to build enormous particle accelerators — the biggest at the present time (1977) might produce 1000 GeV on a good day. Compare 10^{12} eV (1,000 GeV or 1 TeV) with 10^{20} eV! It is not known what process can produce that. Averaged over galactic space, the energy carried by means of cosmic rays is about the same as in starlight.

When a cosmic ray particle hits a nucleus of an atom in the atmosphere a host of secondary particles and γ-rays are produced, and some of these go on to produce still more particles in subsequent collisions. About 100 large air showers a year are produced all over the globe by particles whose energy exceed 10^{19} eV, and in such showers, secondary particles can be detected on the ground over an area of hundreds of metres in radius. Many showers die out before reaching the ground, and only muons, particles like electrons only 200 times heavier, travel down to sea level. These muons are famous for demonstrating the validity of Einstein's prediction that fast-moving clocks go slow. Slow muons spontaneously decay, under the supervision of the weak interaction, into an ordinary electron and a neutrino in a time of the order of a microsecond. Travelling near the speed of light, a muon ought not to get much further than 300 metres before breaking up, but in fact one that is generated in a cosmic ray shower, manages to get through at least 3km of lower atmosphere intact. Its three score and ten years, as it were, has been stretched to seven centuries at least — a remarkable example of time-dilation.

Those cosmic rays which manage to dodge the solar wind, travel fast enough to break through the magnetosphere, and are energetic enough to burst through the atmosphere, either directly or vicariously through their secondary offspring, may at the earth's surface exert a direct cosmic influence on life itself. The delicate chemical copying mechanism at the base of heredity is totally vulnerable to the dispassionate bombardment of cosmic rays. If genetic mutation is essential to evolution and the survival of species, then the long-term wellbeing of life on earth owes a debt to extra-terrestrial forces, as well as to the more mundane influences of chaos and chance. Is it too whimsical to regard our ability to think as the possible consequence of a mysterious event in the galaxy some millions of years ago, signalled by a burst of cosmic rays which coincided with a minimum of the sunspot cycle or, more potently, with a reversal of the earth's magnetic field? — or if not our thinking, perhaps our warm blood, or our lack of reptilian scales, or our adaptations to 'poisonous' oxygen? It is highly unlikely that we shall ever know the exact physico-chemical or biophysical causes of any of these evolutionary steps. Probably there were many processes acting together, but one of them at least, it seems safe to affirm, must have been cosmic.

9.8 NEUTRINOS AND GRAVITATIONAL WAVES

The whole package of cosmic radiation consisting of photons and charged particles seems a thin and bodiless thing when it is compared with the chunky meteoroid flux of the solar system. Yet those photons and charged particles are fat and weighty in comparison with the two other components of cosmic radiation we have yet to mention. These two components are so 'ethereal' that detecting their presence is a major technical feat, and in the case of one of those components, it is by no means sure that it has been detected at all. The two cosmic fluxes we are speaking of are neutrinos and gravitational waves.

Neutrinos are creatures of the weak interaction. They are produced when elementary particles decay into the two stable particles, the proton and the electron. A neutrino is the particle you would be left with if you took an electron and removed its charge and all of its mass, leaving only its spin, like the grin of a Cheshire cat. It is therefore nothing much more than an embodiment of motion, having no mass and no charge. Another way of getting at a neutrino is to take a photon, which already has no charge or mass, but does have spin, and chopping it into two to get half the spin. In short, the neutrino is a lump of spinning nothing that travels at the speed of light. The rest of matter does not want to know about this thing, and so neutrinos can get through enormous amounts without interacting at all. That is why they are so difficult to detect.

Yet neutrinos are real things. In 1930 Pauli proposed that such particles had to exist if the conservation laws of energy and momentum were to survive. In β-ray production, a neutron in a radioactive nucleus turns into a proton and emits an electron plus an antineutrino, or in other types of radioactivity a proton turns into a neutron and emits a positron (antielectron) and a neutrino. Both processes are termed β-decay. Neutrino (and antineutrino) detection depends on the fact that inverse β-decay can take place. Thus an antineutrino ($\bar{\nu}$) may be captured by proton (p) turning it into a neutron (n) plus a positron (e^+); or a neutrino (ν) may be captured by a neutron (n) turning it into a proton (p) plus an electron (e^-). These reactions are summarised as follows:

β-decay	Inverse β-decay
$n \rightarrow p + e^- + \bar{\nu}$	$p + \bar{\nu} \rightarrow n + e^+$
$p \rightarrow n + e^+ + \nu$	$n + \nu \rightarrow p + e^-$

The point is that inverse β-decay provides a way of detecting neutrinos and antineutrinos, even distinguishing between them. Experiments with the immense flux of neutrinos in a nuclear reactor amply confirm their existence. But the probability of inverse β-decay in matter is so tiny that the average distance which a neutrino travels through solid iron before being absorbed is 130 light-years! Our theories of astrophysics, tell us that neutrinos are produced deep inside stars and so may act as a probe into stellar interiors, otherwise inaccessible. Their production means an energy loss to the star, because they are so weakly reabsorbed, and under certain circumstances this may result in a substantial refrigeration which produces total collapse of the star. In such a way neutrinos may be instrumental in creating a supernova, as we mentioned in Section 7.2. Being extraordinarily detached from the rest of us, they can scarcely be described as an active element in our environment. Yet that same detachment gives them a well-travelled look that marks them out as citizens of the whole cosmos rather than provincials within our own galaxy.

Gravitational waves ought to be emitted by pulsating or collapsing stars. Whether this is so or not is still a matter for experimental confirmation. It may be that the earth is tossed gently on a sea of gravitational undulations, or squeezed delicately by shorter wavelengths so that an earthquake is triggered, in which case we will have to admit gravitation waves as active components in the cosmic radiation field: but so far their activity has been limited to the research laboratory, and if they are in some way active in our environment, it is a way which is at present hidden from us. Nevertheless there are good theoretical reasons for their existence, and they ought to have been emitted copiously during the big bang of creation.

SUGGESTED READING

Allen, C. W., *Astrophysical Quantities*, 3rd edition, (University of London, Athlone Press, 1973).

Bahcall, J. N., 'Neutrons from the Sun, (*Frontiers of Astronomy*, July 1969).

Boas, N., 'The Scientific renaissance 1450–1630', (Collins, 1962).

Brandt, J. C. and Maran, S. P., *New Horizons in Astronomy*, (W. H. Freeman, 1972).

Chandresekhar, S., *The Principles of Stellar Dynamics*, (New York Acadamy of Sciences, 1943).

Chapman, C. R., 'The Nature of Asteroids', (p. 24, *Scientific American*, January '75).

Friedman, H., 'X-Ray Astronomy', (*Frontiers of Astronomy*, June 1964).

Goldberg, L., 'Ultraviolet Astronomy', (*Frontiers of Astronomy*, June 1969).

Gribbon, J. and Plagermann, S., 'The Jupiter Effect', (Fontana Books, Collins, 1977).

Grossman, L., 'The Most Primitive Objects in the Solar System', (p. 30 *Scientific American*, February 1975).

Hartmann, W. K., 'The Smaller Bodies of the Solar System', (p. 124, *Scientific American*, Sept. 1975).

Howard, R., 'The Rotation of the Sun', (p. 106, *Scientific American*, April 1975).

Kellerman, W., 'Cosmic Rays; Galactic or Extragalactic?' (p. 249), *Scientific American*, June 1976).

Lewis, J. S., 'The Chemistry of the Solar System', (p. 50, *Scientific American*, March 1974).

Neugebauer, G. and Becklin, E. E., 'The Brightest Infrared Sources', (*New Frontiers in Astronomy*, April 1973).

Parker, E. H., 'The Solar Wind', (*Frontiers of Astronomy*, April 1964).

Pasachoff, J. M., 'The Solar Corona', (p. 68, *Scientific American*, October 1973).

Richardson, R. S., 'The Discovery of Icarus', (*New Frontiers in Astronomy*, April 1965).

Schramm, D. N., 'The Age of the Elements', (p. 69, *Scientific American*, January 1974).

Stephenson, F. R. and Clark, D. H., 'Historical Supernovas', (p. 100, *Scientific American*, June 1976).

Strong, I. B. and Klebesadel, R. W., 'Cosmic Gamma-ray bursts', (p. 66, *Scientific American*, October 1976).

Van Allen, J. A., 'Interplanetary Particles and Fields', (p. 160, *Scientific American*, September 1975).

Webster, A., 'The Cosmic Background Radiation', (p. 26, *Scientific American*, August 1974).

Whipple, F. L., 'The Nature of Comets', (p. 48, *Scientific American*, February 1974).

Wood, K. D., 'Sunspots and Planets', (*Nature*, **240**, 91, 1972).

10
Origins

The infinite and the immeasurable is as necessary to man as the little planet which he inhabits.

Fyodor Dostoyevsky (1821-1881) *The Possessed.*

As for us, we deem the whole world animate, and all globes, all stars, and this glorious earth, too, we hold to be from the beginning by their own destinate souls governed, and from them also to have the impulse of self-preservation.

William Gilberd (1540-1603), *De Magnete.*

10.1 CREATION

Cosmologists play a game about the origins of our busy universe. The basic rules are simple: take as framework the currently most promising theory of the universe, the hot big bang; assume the laws of physics remain unchanged, and apply our prevailing knowledge of thermonuclear processes to describe the evolution of the primeval fireball. Two widely divergent views of the game are (1) that there are no winners, only losers, (2) all players are winners, since it is impossible to lose. The first view reflects the high probability that future knowledge will render cherished stories inaccurate or plainly wrong; a fate, as it happened, which attended the very first idea proposed by that Greek alphabet of cosmologists R. A. Alpher, H. A. Bethe and G. Gamow, the latter being the originator of the game. The second viewpoint expresses the usual optimism of scientists in all fields that their contribution always represents something better than was there before. Whatever the viewpoint, the game is a fascinating one, and were it not for three things which come out of this speculation, it could safely be ignored as a curious metaphysical dialectic on a par with the ancient argument concerning the number of angels who could stand on a pinhead.

The three things are these: (1) the prediction of the evolution of electromagnetic radiation, (2) the prediction of the ratio of the number of helium atoms in the universe to the number of hydrogen atoms, and (3) the affirmation that only hydrogen, helium and possibly lithium, that is, the three lightest elements, were created, and the heavier elements were entirely absent. These are matters which say something about our present environment, and so can be tested. The tale is worth telling (Figure 10.1).

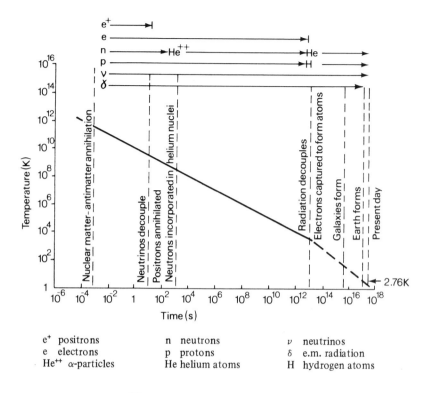

e⁺ positrons	n neutrons	ν neutrinos
e electrons	p protons	δ e.m. radiation
He⁺⁺ α-particles	He helium atoms	H hydrogen atoms

Figure 10.1 – History of Universe

In the beginning there was energy, temperature and expansion. The energy manifested itself in all types of elementary particles and their motions, and the temperature was about 2×10^{12}K. A millisecond later, (please note the time scale here – not centuries, not even years, but milliseconds!), cooled below 10^{12}K by the expansion, nuclear matter and antimatter annihilated each other and the universe became an expanding ball of neutrinos, antineutrinos, electrons, positrons, muons, antimuons, gamma rays, and those few protons and neutrons which escaped annihilation, either because of a basic asymmetry between the amounts of matter and antimatter, or because of a physical separation of matter and antimatter. For ten seconds matter remained in the form primarily of leptons (neutrinos, electron, muons and their antiparticles) interacting rapidly with one another, but then the temperature fell below 10^{10}K, and many changes took place. Neutrinos, once and for all time, became almost entirely decoupled from both matter and radiation, and began their own individual, ghostly evolution. Electrons annihilated positrons and each annihilation created two gamma-ray photons. But the number of electrons which remained exactly equalled the number of protons, in order to preserve electrical neutrality, and so once again the asymmetry between matter, (in the form of electrons) and antimatter (in the form of positrons) was manifested. Once these processes were complete, protons and neutrons combined to form first deuterium, the heavy isotope of hydrogen, and then helium in the form of alpha particles. After twenty minutes this process was complete and all neutrons had been incorporated into helium nuclei, except for a very few which had become deuterium or lithium nucleons. For every twelve protons there was one alpha particle, and that is the present day ratio of hydrogen to helium atoms observed throughout the universe. Radiation and matter continued to interact strongly until the expansion, after about a million years, cooled the universe to below 10^4K. Thereafter, the interaction became weak, and matter and photons went henceforth on their separate ways. About that time, radiation received a tremendous boost when electrons combined with protons and alpha particles, emitting light as they did so, to form neutral hydrogen and helium atoms. Shifted in frequency by the Doppler effect associated with the continual expansion, this radiation, 10^{10} years later, is the cosmic microwave background we observe today. A thousand million years passed, and galaxies of stars began to form out of the primordial hydrogen and helium, and to evolve into the universe as we know it now, complete with its planets and life-forms.

It is a remarkable history, more myth than superstition, more science than myth (to emulate Yeats); with the power of myth to stir the blood, having yet the unique strength of science to explain and predict. As far as the cosmic neutrino flux (does it exist?) is concerned, the world was created in ten seconds. Primordial hydrogen and helium would have that time as being twenty minutes, and electromagnetic radiation must argue for a creation stretching over a million years. But earth has a gravitational strength too weak to retain the cosmic proportions of hydrogen and helium, and it encounters neutrinos but rarely. The only creature of primal creation which the earth experiences is the microwave background, an echo of that great fall of electrons into atomic bondage after a brief million years of paradisiacal freedom.

10.2 THE STARS

But we cannot explain everything by a big bang theory on its own. The present day γ-ray, x-ray, ultraviolet, visible, infrared and radio backgrounds remain unexplained. The theory says nothing about the elements heavier than lithium; nor does it describe the origin of planets and of life. Such matters are connected with the idiosyncratic behaviour of galaxies and of individual stars, rather than with the cosmos. They are provincial details in comparison with the creation of neutrinos, photons, hydrogen and helium; a matter of dotting i's and crossing t's. If we are to concern ourselves with such minutiae, we had better look to astrophysics rather than to cosmology in order to put our immediate physical environment in historical perspective.

It is true that some of the radiation received by earth can be traced to explosive events involving whole galaxies. Such may have produced the intense radio and x-ray emission of so-called Seyfert galaxies, or the radio and optical emission of quasars, those quasi-stellar objects thought to be components of distant clusters. But the bulk of radiation falling on the earth originates in stars of our own galaxy, and it is to stars in our own galaxy that we look for the origin of the rest of the elements, for the birth of the solar system, and for life itself.

We must therefore make a leap in space, as well as the leap in time we have already made since the creation of the universe. From the universe at large, some 10^{10} light-years across and expanding, with its clusters and superclusters of galaxies, we must come down to the comparatively tiny galactic atom, the star. The scale has to contract through the 10^7 light-years of an average-sized galactic cluster with its 130 galaxies, contract further past the 10^5 light-years of an average galaxy with its 10^{11} stars and its gas clouds, and diminish by a factor of a hundred thousand to a single light-year, before we can regard ourselves as being in the neighbourhood of a single star. Even then, we need a contraction in scale by yet another factor of a hundred thousand to reach the dimensions of its 'solar system' of planets, if it has one. Diminishing from the size of the galaxy to the size of the earth's orbit round the sun involves a contraction factor of 10^{-10}, the same factor relating the size of a piece of furniture to one of its atoms, or the size of the earth's orbit to a room-full of furniture (Figure 10.2).

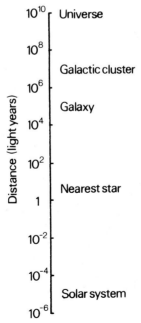

Figure 10.2(a) — Distance scale.
(1 light-year = 9.46×10^{15}m)
Figure 10.2(b) (*following page*)
Forms of galaxies (Hubble's classification of spiral nebulae).
(Science Museum, London)

Sa NGC 4594

SBa NGC 28

Sb NGC 2841

SBb NGC 58

c NGC 5457 (M101)

SBc NGC 74

A star is minute compared with the galaxy, the galaxy trivial in comparison
with the universe. As a fireball, the star does not begin to compete with the
primordial big bang. The latter predestined the universe firmly to be a place in
which on average there were in every cubic metre (1) an energy of some 10^{-11}
joule tied up as hydrogen and helium with traces of deuterium and lithium (2) an
energy of some 10^{-14} joule as microwave photons, and (3) an energy of some 10^{-14}
joule in the form of ghostly neutrinos (and possibly even ghostlier, gravitational
radiation). All stars can do is fool around a little with the original matter and
add traces of the heavier elements, a few photons at other wavelengths, and a
dash of neutrino flux to the cosmic mixture.

Yet importance, like beauty, is in the eye of the beholder, and to us on our
sorely dependent planet, the sun's fooling about, insignificant though it may
appear to be on the cosmological scale, seems not without significance to us.
From the standpoint of intergalactic space the contribution of the sun, indeed
of all the stars in the universe, is minute. Within our galaxy, in interstellar space,
things are already very different from the cosmic average: the average matter
density is increased by a factor of some 10^7, and starlight and cosmic rays reach
the level of the cosmic microwave background. Our planet is bathed in sunlight
whose energy density is about 4×10^{-6} joules per cubic metre – four hundred
million times as intense as the microwave background. That figure of four hundred million is the measure of how much the sun insulates us from the bleak
intergalactic vacuum. What stars are, what they do, and how they evolve are
therefore questions of considerable interest to us. Eternals may have been defined
long ago in the holocaust of creation, but we are transient, evolving beings,
veritable creatures of the sun, and are naturally and prudently concerned with
the life-style and life-cycle of our local star and its fellows in the galaxy, all of
them ephemeral like ourselves.

10.3 STELLAR EVOLUTION

A star is born out of gravitational contraction of a huge gas cloud. Such
clouds inhabit parts of our galaxy and manifest themselves (through telescopes)
as masses of dark or faintly glowing nebulosities, obscuring the stars lying
behind them. Wherever they exist, stars are being born. (Figure 2.1(b), page 29).

Interstellar clouds consist mainly of hydrogen – about eight atoms and one
molecule in every cubic centimetre – plus traces of other atoms and molecules,
some of the latter being organic. At anything from a hundred to ten thousand
sites within the cloud, protostars form. A local concentration of matter attracts
more matter by gravitational attraction, and as the atoms and molecules free-
fall they pick up speed, collide with one another, and generally acquire that
chaotic motion which we call heat. Gravitational contraction therefore causes
the temperature of the gas to rise. Such is the tenuousness of interstellar gas
that the process of star formation takes about 10^7 years (Figure 10.3).

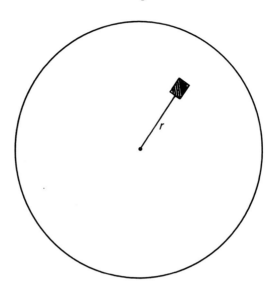

Equation of motion of element of protostar.

$$\rho \frac{d^2r}{dt^2} = -\frac{dP}{dr} - G\frac{\rho M_r}{r^2}$$

M_r = mass within radius r, $\dfrac{dP}{dr}$ = pressure gradient.

In the early stages of formation the pressure force is negligible.

If $M_r = \dfrac{4\pi}{3}r^3\rho$ (uniform gas cloud)

$$\frac{d^2r}{dt^2} = -\frac{4\pi G\rho}{3}r \text{ (simple harmonic motion)}$$

$\therefore$ Characteristic time $\approx \dfrac{1}{\omega} = \left(\dfrac{3}{4\pi}G\rho\right)^{\frac{1}{2}} \approx 10^7$ years.

Figure 10.3 — Formation time of star.

Whether the star is stillborn, normal or unstable depends upon the mass of gas involved in the condensation. A useful measure is the mass of the sun, 1.99×10^{30} kilograms, symbol M_0. If the mass is less than about 0.1 M_0 there is insufficient gravitational muscle to get the temperature up to 10^7K, the temperature required to fuse hydrogen and produce helium. If the mass exceeds about 100 M_0, there is so much gravitational muscle that radiation inside the star, once nuclear fusion has been triggered off, becomes intense enough for radiation pressure to be more important than ordinary gas pressure. Under these circumstances the star becomes hydrodynamically unstable and has to explode matter to achieve normality. Normal protostars thus have masses between the rough limits 0.1 M_0 and 100 M_0. Once the temperature at the centre reaches 10^7K, the nuclear energy production caused by the fusion of protons to form helium nuclei halts gravitational contraction, and the star becomes what is termed a 'main-sequence' star. As a main-sequence star it is characterized by a hydrogen 'burning' core, a stable radius, and an output of radiation energy equal to the energy released by the conversion of hydrogen into helium. Such a star is our sun (Figure 10.4).

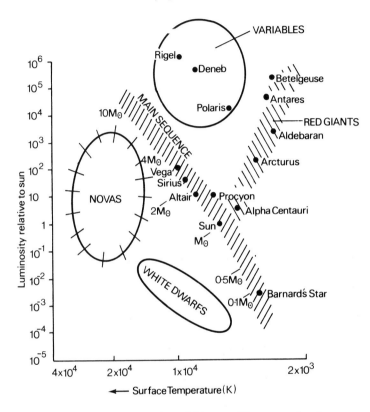

Figure 10.4 – Herzsprung-Russell Diagram. (M_θ = mass of sun)

How long a star remains on the main sequence depends on how long it takes for hydrogen at the core to be used up. The more massive the star, the quicker is the hydrogen used up. A star of mass 15 M_0 will remain on the main sequence a mere 10^7 years, which is 460 times shorter than the age of the earth. That we see such stars in the heavens is evidence that stars continue to be born within our galaxy. On the other hand, the sun is expected to remain on the main sequence 8×10^9 years. If the ages of the earth and the sun are the same, this means that the sun ought to have another 3×10^9 years of placid main sequencing to look forward to. If, however, the sun acquired the solar system when it was already on the main sequence, as some theories have postulated, there is no knowing how long the sun will last in its present form.

What happens when the central hydrogen is used up? As the proportion of helium increases it is thought that the star becomes hotter and more luminous. At some stage the hydrogen at the core is entirely replaced by helium. The detailed behaviour after this crucial point is reached depends very much on how much mixing there has been, and this in turn depends upon mass – the greater the mass, the more the mixing. Generally, once a helium core is formed, no further nuclear reactions can proceed at the centre, and so the core has to contract in order to maintain a sufficiently high temperature in the skin around the core for hydrogen burning to continue. Once the helium core is above a critical mass it becomes mechanically unstable and contracts rapidly and heats up, and simultaneously the outer layers expand and the surface cools. The star becomes a red giant, with a radius fifty to a hundred times that of the sun. A hundred-fold expansion of the sun would bring its surface out beyond the orbit of Venus, and half way to the earth. The latter would have become uninhabitable long before red-gianthood, at about the time the central hydrogen became significantly depleted.

In any case, a red giant does not make a comfortable neighbour. Deep inside, there sits a potential helium bomb. If the mass of the star exceeds about $0.4 \, M_0$ the central temperature, through contraction, will ultimately reach some $10^8 K$, and helium will burn (in the nuclear sense) to carbon. When this happens the star is thrown into severe convulsions during which considerable amounts of matter may be blown off into interstellar space. Once helium is used up at the centre, a carbon core forms and the star undergoes another phase of evolution which leads to the red giant state, though with greater luminosity than before. If sufficiently massive, the star can achieve a high enough temperature for carbon to burn, with once more an explosive loss of mass, and go on repeating the cycle with other nuclear fuels, until an iron core is built up. Thereafter no further source of energy is available except by gravitational contraction. It is thought that ultimately, at temperatures around $7 \times 10^9 K$, various nuclear and elementary-particle interactions occur which either absorb energy or convert energy into neutrinos which easily escape. Refrigeration of this kind produces a catastrophic collapse, which itself would rapidly heat up the outer layer, trigger hydrogen and helium fusion, and blow a substantial part of the star to bits, incidentally creating a number of the elements heavier than iron. Such is thought to be the cause of a supernova, and the origin of the heavier elements.

Once all nuclear fuel is used up, the star, or whatever is left of it after its evolution, cools. Again its mass is crucial for determining its doom. If its mass is less than about $1.4 \, M_0$ it becomes a white dwarf, cooling gradually into a black dwarf, some hundred times smaller in diameter than the sun. A star whose mass is above $1.4 \, M_0$ cannot find enough outward pressure-force within it to withstand gravitation, and so it contracts. A stable end-point, nevertheless, can be reached for masses less than about 2 M_0, when the density reaches that of nuclear matter (some 10^5kg m^{-3}). Under these conditions electrons are forced into the nuclei of atoms to produce neutrons, and the end-point is a neutron star, about 100 kilometres in radius. These objects are very plausibly identified with pulsars, tiny objects which emit periodic bursts of radio waves with periods as short as 30 milliseconds. If the mass exceeds 2 M_0 nothing can stop the star contracting indefinitely to produce a, so-called, black hole — an object whose gravitational field is so large that even light cannot escape from its surface.

The more massive the star, the faster it evolves. This has interesting consequences for stars which are gravitationally bound together in binary systems. The more massive component evolves the quicker, and reaches the red giant phase faster, then its less massive companion. The expansion into a red-giant may result in a spectacular transfer of matter from the more massive to the less massive star, and this may be the cause of a nova.

It would seem that the sun, not being a member of a binary system, and having a modest mass, may look forward to a reasonably quiet future with a dwarfish retirement — reasonably quiet, that is, compared with what can happen. There is no doubt, however, that terrestrial conditions will become intolerable once the sun begins to progress from its stable main-sequence behaviour towards being a red giant.

Probably this will not happen for a thousand million years — time enough for tens of thousands of technologically based societies to have risen and fallen — time enough to plan the escape route of life on earth. But we cannot be sure. We think the earth and the solar system came into being 4.6×10^9 years ago. Did the sun also? Were sun and planets formed out of the same nebula at the same time? We are aware that all that vast amount of matter we call the Milky Way Galaxy did not condense straight away into stars as soon as the galaxy itself condensed, for star-formation goes on today. Some of the stars we see in the sky, were indeed first to form. Most, however, have formed later than the initial galactic condensation, among them the sun. When did the planets form and how did the solar system come about are questions which have no completely definite answer at the present time. There are, nevertheless, no shortage of guesses.

10.4 THE SOLAR SYSTEM

There are certain peculiar characteristics about the solar system which have to be explained by any theory of origin. Among these, three stand out as being crucial (Figure 10.5).

Figure 10.5 — Components of the Solar System (Geological Museum, London)

1. The total angular momentum of the system is distributed in a very odd way. Angular momentum is a measure of the amount of rotational motion, defined quantitatively as the momentum (mass times velocity) multiplied by the distance of the line of momentum from the centre. Thus a rapidly moving body rotating in an orbit of large radius has a high angular momentum. A body like the earth which spins about its own axis has rotational angular momentum as well as orbital angular momentum. We might expect that the sun, having most of the mass of the solar system, would also, by its rotation, have most of the total angular momentum. Exactly the opposite is the case. The planets and their satellites, whose mass is only 0.14% of the total mass of the solar system, account for 98% of the total angular momentum by their rotation and orbital motion. If all the observed angular momentum were indeed in the sun, the sun would rotate some fifty times faster than it does. From the standpoint of simplest expectation, the sun rotates far too slowly.

2. There is a distinctive distribution of matter in space. All planetary orbits are roughly coplanar, and the planets themselves fall into two categories, the minor planets, small and dense (Mercury, Venus, Earth, Mars) and the major planets, large and of low density, (Jupiter, Saturn, Uranus, Neptune; but Pluto does not seem to fit this scheme). The average distances, r_n, from the sun obey a law first discovered by Bode in 1772:

$$r_n = 0.4 + 0.3 \times 2^n \text{ A.U.}^\dagger$$

where $n = \infty$ for Mercury, $n = 0$ for Venus, $n = 1$ for Earth, $n = 2$ for Mars, $n = 3$ for asteroids, $n = 4$ for Jupiter, and so on (again Pluto is something of a maverick, and goes along with Bode's Law only approximately). That Bode's Law is not just a numerical curiosity is a view supported by the fact that the inner satellites of Jupiter, Saturn and Uranus obey laws of a similar form.

3. Roughly, the abundance of chemical elements observed spectroscopically in the sun's surface agree with those found on earth, apart from the light elements hydrogen and helium which are not retained by the earth's gravity, and lithium, beryllium and boron, which have a low solar abundance because they are destroyed by nuclear reactions at temperatures of the order of 10^6K. This suggest that the earth has never been that hot. A low temperature for formation is also suggested by the fact that the volatile elements such as cadmium, zinc and mercury are not highly concentrated at the earth's surface, as they would have been had they been vapourized at some stage.

The distribution of angular momentum, the distribution of matter, and the distribution of chemical elements are the basic properties of the solar system which have to be explained by a theory of origin.

---

†Astr. A.U. = Astronomical Unit (1AU = mean earth-sun distance).

Theories may be divided into two classes: those in which it is assumed that the sun and planets had a common origin; and those which assume that the planets were acquired after the sun was formed. In the first class the sun and planets condense out of a largely gaseous nebula. Nebula theories are attractive in their economy of providing a planetary system as an almost literal spin-off of star formation. They have the disadvantage, on the other hand, of having to invent intricate explanations for the angular momentum distribution. In the second class, planets are acquired in one of two ways: collision with another star, or accretion of material from an interstellar dust and gas cloud. Since in both collision theories and accretion theories the sun moves relatively to the source of planetary material, that material once captured, possesses a large angular momentum. Explaining the observed distribution of angular momentum is therefore no problem.

The first nebula theory was that of Laplace in 1796. A rotating gas nebula contracted under gravity. Since angular momentum is conserved, the angular velocity increases and centrifugal forces reduce the spherical cloud to a disc, and eventually rings of matter are thrown off and each ring subsequently condenses into a planet. Laplace's theory is not tenable because it leaves the sun rotating too quickly; neither is the observed total angular momentum sufficient for matter to be ejected as envisaged. In 1960, Hoyle got around these difficulties by starting with enough angular momentum for centrifugal forces, eventually, to eject a ring of material at about the radius of Mercury's orbit. He argued that a magnetic field, interacting with the ionized gas in the ring, would transfer angular momentum from the central condensation, the sun, to the material in the ring, which would spiral outwards and ultimately condense into planets, or would escape, carrying the excess angular momentum away. Another way out was taken by von Weizsacker in 1944. He postulated that turbulent motion in the contracting disc would produce whirling vortices rotating as proto-planets around the inner condensation. This would acquire only a small amount of angular momentum, because only material which was not whirling would be captured. Excess angular momentum has to be lost by matter escaping from the system. More recently ter Haar (1950) and Kuiper (1951) have developed von Weizsacker's idea.

Yet another approach is McCrea's floccule theory (1960), that the initial condensation of a large interstellar cloud, from which hundreds of stars ultimately form, was comprised of small, planet-sized 'floccules'; some of which coalesced to form stars; other being captured to form planets. A body of density ρ would be disrupted by gravitational tides raised by the sun if it were nearer than the Roche limit given approximately by $1.5 \ (M_0/\rho)^{1/3}$. McCrea's average density of floccules throughout the cloud would lead to this distance being of the order of Jupiter's orbit. Thus giant planets could form by coalescence at this distance and beyond, but not nearer, thus leading to a division between major and minor planets. As in collision and accretion theories, the distribution of angular momentum is made extremely plausible, but it is difficult to account for Bode's Law.

Collision theories assume a sun, already formed, colliding with another star. A well-popularized model advanced by Chamberlain (1901) and Moulton (1905) and developed by Jeffreys and Jeans between 1916 and 1919, obtained planets from a huge cigar-shaped tide pulled out of the sun by a passing star. This entails a hot origin and we know that it conflicts with the chemical evidence. Nevertheless, a collision theory is attractive because it explains simply the angular momentum distribution. But it should involve cool matter, and such a collision theory was revived by Woolfson in 1964. This time the tide was pulled from out of the colliding star, seen as a diffuse protostar, not yet fully formed and hence still cool. If we have remarked in Chapter 9 that collisions between stars are highly improbable, that was assuming the density of stars to be as observed in the neighbourhood of the sun. In a gas cloud, in which hundreds of stars are in the process of being born, collisions between protostars, and between fully-formed stars and protostars, would be more frequent, so Woolfson's model is plausible. The matter in the tide of stellular gas is pulled entirely away from the protostar and is captured into orbit around the sun. In this view, the planets arise from a different condensate of the same nebula — the colliding protostar — some time later than the birth of the sun, but probably no later than 10^8 years, the time of formation of stars less massive than the sun.

Accretion theories assume that the sun captures material into orbit as it passes through an interstellar gas and dust cloud — perhaps the debris of a supernova. Schmidt, who proposed the theory in 1944 and developed it in 1959, had to assume the presence of a third body — a nearby star — in order to get any matter captured at all, otherwise conservation laws of energy and momentum forbade it. To assume this highly improbable situation is however unnecessary, according to Lyttleton, who showed in 1961 that if the sun's velocity relative to the cloud were as low as 200 metres per second, enough material would be captured. Accretion theories share with collision theories the avoidance of the problem of the slow solar rotation, although they explain satisfactorily the relatively high angular momentum associated with the planets. How the remarkably slow rotation of the sun comes about is left unanswered. And, of course, there is no hint in any accretion theory as to how late in the sun's life the planets were acquired.

All theories have the problem of explaining the nearly circular orbits of the planets and Bode's Law. Explaining circular orbits is difficult, especially for collision and accretion theories, and no generally acceptable theory for Bode's Law exists. All theories have to worry about the division into major and minor planets. The common explanation is to assume, very plausibly, a temperature gradient whereby the material nearest to the sun is hotter than the more distant. Stoney material with a high melting point, solidifies first — even the major planets may have dense stoney cores – but the more volatile, light materials like hydrogen, helium, ammonia, etc. escape from the hot region, where the minor planets form, and subsequently aggregate as the low-density major planets.

Obviously, the great number of uncertainties present makes it impossible to be definitive about our planet's origin, and how frequent such an occurrence may have been in the galaxy. There is substantial astronomical evidence that dark companions to other stars exist, and the slow rotation of low mass stars like our sun, could be an indication of planetary systems. The balance of evidence is in favour of the existence of solar systems elsewhere in the galaxy, but without a firmly convincing model of origin it is not possible to make an accurate assessment of number.

10.5 LIFE

A cold origin of the earth is implied by the study of the abundances of chemical elements, and the general study of the chemistry of the solar system as carried out by Urey and others. Nevertheless, during the contraction under gravity of matter which formed the earth, high temperatures of order 2000K would be developed. Such a temperature would be high enough to melt silicates, and to allow hydrogen to reduce oxides of iron to iron. Iron plus a few other elements would sink to the centre of the earth, and the molten basaltic liquids would rise. Eventually such processes would produce the iron core, the mantle, and the lighter silicate layers near the surface, which is the internal structure of the earth.

The original atmosphere was almost certainly a reducing one, in which free oxygen was practically entirely absent, and hydrogen much more abundant. Steam, ammonia (NH_3), methane (CH_4), carbon monoxide (CO), carbon dioxide (CO_2) and nitrogen, in addition to hydrogen would make for an extremely noxious atmosphere, and given the high temperature of the earth at the time, a very turbulent one. There would be an abundance of energy to break up simple molecules and, as the earth cooled, to form complex ones; to break these down to simple ones again, or form even more complex molecules. In that cauldron of the original atmosphere, many possible combinations of these elements pertinent to life – hydrogen, oxygen, carbon and nitrogen – must have to come into existence, sometimes under the transient shock of a lightning stroke. Laboratory experiments confirm that electric discharges passing through a gas mixture simulating the primitive atmosphere produce amino acids, which is the stuff of proteins. If simple organic compounds were formed in this way they would be destroyed immediately if oxygen were present. Oxygen is basically poisonous; it destroys organic things. So oxygen was almost certainly absent as free atoms or molecules. As the earth cooled below the boiling point of water, the atmospheric steam would have condensed and filled up the hollows of the irregular surface of the earth, so forming the seas, lakes and oceans. Organic material in the atmosphere would have eventually been washed by rain out of the gaseous environment into the more concentrated aqueous ambient of the oceans.

Biochemists speculate that, as the concentration of organic molecules grew, coagulations of organic material in groups would occur. A group would grow by the addition of other organic molecules (which began to act as food for the clusters); and ultimately it would get too big, and break in two or more bits, which would themselves begin to grow. Eventually a system evolved which, when it split up, produced two replicas of itself. That system would be alive, and given a sufficient food and energy supply, would thrive and reproduce. That happening would mark the beginning of viable life on the planet. The impressive uniformity of organic material found in all forms of life today points to identical ancestors. Perhaps, once the concentration of organic molecules in the oceans reached a critical level, a certain assembly of molecules (which had hitherto led a tenuous, intermittent existence) suddenly or gradually found things to their liking. Their numbers grew into a 'population explosion', and exploded at the expense of the other species. These primitive structures would give rise to other forms of life by evolution. The microfossils found in rocks 2.6×10^9 years old, are fossils of one-celled creatures which probably had developed photosynthesis, that is the use of sunlight to convert carbon dioxide and water into glucose and oxygen. Others resemble fossilized bacteria. Such animals were well-advanced compared with their primitive forebears. A cell, after all, is already a highly organized community containing, within its few microns diameter, many small living structures that have learnt to get along with one another. Presumably, those subcell structures represent earlier individual systems which found survival easier together than apart; rather as if a few viruses of one sort and another had decided to form a community and settle down.

The advent of a cell some three thousand million years ago which could photosynthesize glucose was poison to the rest. Like dirty chemical factories, huge quantities of them emitted deadly oxygen and changed the atmosphere gradually to the composition at the present day. All cells were developed in a reducing atmosphere. They had to learn to cope with oxygen or die. Some, the anaerobic bacteria, may never have learnt and were compelled to limit their habitat to airless bottoms of ponds. The rest of us cannot do without it.

Just as subcellular structures banded together to form a cell, cells banded together to form organisms of all sorts, and organisms banded together to form packs or tribes, and ultimately man formed nations, and groups of nations. A supernova spawned a nebula, the nebula spawned the earth, the earth spawned a biosphere – an animate thing born out of the physical environment of long ago. To Lovelock and Margulis (1974), the biosphere is a living entity they call Gaia (after the Greek goddess of earth), able to control temperature, atmospheric and oceanic composition and materials on the earth's surface. To Thomas (1974) the earth is somewhat like a single cell of immense complexity. The biosphere and the physical environment become one. Indeed, in this view, animate and inanimate merge; and the earth becomes a living entity feeding off the sun, possessing a limited self-conciousness, but aware of the immensity of the surrounding galaxy and the dark expansion of the universe.

10.6 LIFE IN THE GALAXY

Such speculation as the Gaia hypothesis encourages us not to regard the traditional division between the physical and the biological as inviolate, especially at the planetary level. If the physical processes which formed the solar system had in them the potentiality for producing life, realised in the case of the earth, it is pertinent to include, in the investigation of the physical environment, the study of the possible occurrence of extra-terrestrial life-forms. In Chapter 7 we saw that for chemically-based life to exist, the temperature had not to be so low that important chemicals like carbon dioxide and ammonia were frozen, yet not so high that complex organic molecules dissociated rapidly. With a very generous interpretation of favourable conditions, that temperature range stretched from 200K to 6000K. A more realistic range may turn out to be much narrower, say from 250K to 750K. In either case, appropriate thermal conditions exist in the solar system only on the minor planets.

The maximum steady temperature a planet could achieve, relying on the sun's radiation, is if it were to behave like a blackbody, absorbing all solar radiation falling on it, and radiating away into space its heat as infrared radiation. If a planet is regarded as a perfectly conducting sphere absorbing a fraction, say 0.65 of the sun's energy, and radiating like a blackbody, we can calculate its temperature, using the known luminosity of the sun, the inverse-square law, and Stefan's Law. For those planets with an atmosphere, the surface temperature will be enhanced by at least a factor of $2^{\frac{1}{4}}$ (Section 8.5). On the basis of such a calculation (Table 10.1) only Mercury, Venus, Earth and Mars could develop chemical life. Of these, Mercury looks utterly dead, and at the time of writing so does Mars, though in the latter case the 'look' *via* unmanned spacecraft, has been considerably more searching. Venus remains mysterious, and so do the atmospheres of Jupiter and Saturn. It is possible that life has developed only on the earth.

For life elsewhere in the galaxy to be observable it surely must have developed a society materially based on science and technology, otherwise how could we observe it or interact with it. Unless one regards the physical origin of the solar system as having been the result of a rare, highly improbable, set of circumstances, or that the origin of life is equally as improbable, then one must accept that the galaxy swarms with life. The question then is, to what degree?

Of course, no reliable answer can be given. The uncertainties are too numerous. At most we can identify the factors which enter the problem, bewail our ignorance of the physical, biological and sociological processes which are fundamental to the understanding of the problem, and settle for informed guesses at the upper and lower limits for the number of technical civilizations. There is, at the present time, at least one such civilization — our own. If every star supported a technical civilization there would be some 10^{11} in the galaxy. Thus we can say straight away that the number — let us call it N — lies between 1 and 10^{11}. Can this enormous range be narrowed down?

Table 10.1 — Temperatures of the Planets

Planet	Distance (m)	Temperature Calculated (K)	observed (K)	Minimum surface Temperature[a] (K)
Mercury	5.85×10^{10}	402	$611^{(b)}$	402
Venus	1.08×10^{11}	296	750	352
Earth	1.50×10^{11}	251	300	299
Mars	2.28×10^{11}	204	270	243
Jupiter	7.80×10^{11}	110	135	131
Saturn	1.45×10^{12}	81	125	96
Uranus	2.92×10^{12}	57	103	68
Neptune	4.58×10^{12}	45	108	54
Pluto	6.00×10^{12}	40		47

(a) Assuming thin-atmosphere 'greenhouse' enhancement of $2^{\frac{1}{4}}$ for all but Mercury.
(b) Sun-facing hemisphere.

A star takes tens of millions of years to form and then, if conditions are right, let us suppose a technical civilization appears in some 10^9 years. Because the age of the galaxy must be some 10^{10} years, ten times longer than the growth time for civilizations, the rate of appearance of technical civilizations must have equalled the rate of appearance of suitable stars some 10^9 years ago. The rate of generation of stars, whether suitable or not, is a physical quantity we can estimate roughly from the fact that a star lives some 10^{10} years and there are 10^{11} stars in the galaxy. If this number represents the steady state, which is reached when as many stars die each year as are born, then an average lifetime of some 10^{10} years implies a generation rate between 1 and 10 per year. If each star produced a technical civilization, the latter would be the present day generation rate of technical civilizations. It is surely unlikely that this would be the case. Not all stars develop planetary systems, and the more massive may evolve too rapidly for life to evolve sufficiently. Even if planets existed, life might not develop for some reason or other, or if it did, it might never develop at the intellectual level necessary for science. Such factors must certainly reduce the generation rate by a factor of ten at least. Thus the maximum rate at which advanced civilizations emerge could be no more than about one a year. Estimates of this rate, which have been made by astrophysicists such as Sagan and Shlovskii and others, range from 1 to about 10^{-2} per year. The fact of the matter is — nobody has any firm idea.

What about the average lifespan of a technical civilization? Nobody knows anything about this either. Our own society, as a galactic entity dates from — when? — the development of radio? — at any rate, a matter of tens rather than hundreds or thousands of years. But it is enough to have revealed a number of dangers to which a technological community is exposed — nuclear and biological warfare, (pollution and over-population, may deplete but do not eliminate). The lifetime of an advanced community could be as short as a few hundred years; its memorial a sterile, dusty planet. Or it could last as long as the evolution of the sun allows — a matter of some 10^{10} years. Here is the greatest uncertainty by far. What is the average lifespan of a highly developed, technological community? To ask the question of present-day sociology is like asking Thales of Miletus, the first philosopher, what is the half-life or uranium 238. Estimates of the average lifespan of a civilized community have tended to range between 10^4 and 10^7 years. The characteristic time for stellar evolution involving thermal changes induced by gravitational contraction is of the order of 10^7 years, and so roughly is the evolutionary time-constant. After 10^7 years perhaps, the society evolves away from the scientific-technological society to something quite different. Perhaps it perishes like the dinosaurs. Groping in ignorance, we could probably do worse than think of 10^7 years as a plausible upper limit, rather than 10^{10} years, for the lifespan of a society. And, being marginally pessimistic, we might take the lower limit to be 10^4 years.

Multiplying rate by lifetime gives the steady-state number of technical civilizations in the galaxy. The maximum number is obtained by taking the maximum rate of one new civilization a year and multiplying by a lifetime of 10^7 years to give 10^7. The minimum number is obtained by taking the minimum rate of 10^{-2} a year and multiplying by a lifetime of 10^4 years, to give 10^2. With many qualms, then, we may guess that at any time throughout the galaxy there are between a hundred and ten million technologically based civilizations.

We can use the average density of stars to estimate the average distance separating planetary communities. If there are only 100 in the whole galaxy, each is separated from its nearest neighbour by about 4,000 light-years; but if there are ten million the separation is only about 100 light-years. Even 100 light-years is a considerable distance. If electromagnetic radiation is indeed the fastest carrier of information in the physical world, the gap of 200 years between question and answer is not going to make for witty repartee, to say the most, much less to allow ordinary dull propaganda and tub-thumping. At present, lacking the hyperspace drive and instantaneous transmission of science fiction, we cannot see that the presence of a civilization like ours, a hundred or a thousand light-years away, can be anything but an intellectual curiosity, even if it does exist. Perhaps what may give us pause is the thought of the endless possibilities to which the process of evolution can give rise; because these possibilities might surely have become realities somewhere in our immense galaxy. This evocation of a strand of ancient Greek philosphy — Plato's assertion that every possible intellectual form has its material correspondent — leads to the sobering thought that we may even yet be at a too primitive state to appreciate the nature and power of the highly evolved cultures of the galaxy. If everything possible that can evolve has evolved somewhere, an even more sobering thought is that it may be so different from what we imagine, or can imagine, that its presence could be all around us, and yet we could fail to recognize it.

SUGGESTED READING

Allen, C. W., *Astrophysical Quantities*, 3rd edition, (University of London, Athlone Press, 1973).

Curtis, H.,*Biology*, (Worth, New York, 1968).

Davidson, W. and Narlika, J. V., 'Cosmological Models and their Observational Validity', *Rep. Progr. in Phys.*, **29**, 539 (1966).

Dyson, F. J., 'Energy in the Universe' (*New Frontiers of Astronomy*, September 1971, Freeman).

Gamow, G., 'The Evolutionary Universe' (*Scientific American*, September 1956).

Hoyle, F., *Frontiers of Astronomy* (Heinemann, London 1970).
                Astronomy and Cosmology (W. H. Freeman, San Francisco 1975).

Huang, S., 'Life Outside the Solar System' (*Frontiers of Astronomy*, April 1960).

Lovejoy, A. D., The Great Chain of Being, (Harvard University Press, 1936).

Lovelock, J. E., 'Thermodynamics and the Recognition of Aline Biospheres', *Proc. R. Soc.*, **B189**, 167 (1975).

Narlika, J., *The Structure of the Universe* (Oxford University Press 1977).

Pirie, N. W., *et al.*, 'A discussion on the recognition of alien life',*Proc. R. Soc.*, **B189**, 137-274 (1975).

Rose, W. K., *Astrophysics*, (Holt, Rinehart and Winston, N. York, 1973).

Sagan, C. and Drake, F., 'The Search for Extra-Terrestrial Intelligence', (p. 80, *Scientific American*, may 1975).

Sagan, C., *The Cosmic Connection,* Hodder and Stoughton, London, 1974).

Sciama, D. W.,*Modern Cosmology* (Cambridge University Press, 1971).

Tayler, R. J., *The Stars: their Structure and Evolution*' (Wykeham, London 1972).

Tayler, R. J., 'Origin of the Elements', *Rep. progr. in Phys.*, **29**, 489 (1966).

Weinberg, S. *The First Three Minutes*, (Andre Deutsch, London, 1977).

Woolfson, M. M., 'The Evolution of the Solar System', *Rep. Progr. in Phys.*, **32**, 135 (1969).

Young, J. Z. 'An Introduction to the Study of Man' (Oxford University Press, 1971).

Problems

Proposition – To find the value of a given examiner.
Example – A takes in ten books in the Final Examination, and gets a 3rd Class: B takes in the Examiners, and gets a 2nd. Find the value of the Examiners in terms of books. Find also their value in terms in which no Examination is held.

Lewis Carroll (1832–1898)
Dynamics of a Particle

PROBLEMS CHAPTER 2

2.1 Imagine an infinite, flat-space universe consisting of opaque spheres, cross-sectional area A, uniformly distributed with a concentration N per unit volume. Prove that the sky appears from earth to be completely covered with spheres if the radius R of the universe exceeds $3/(NA)$.

2.2 If the average density of galactic material is 10^{-27}kg m^{-3} and the average number of stars in a galaxy is 10^{11} what is the concentration N (a) of stars, (b) of galaxies, assuming that the sun is an average star? (Mass of sun = 1.99×10^{30}kg.). [**Ans.** (a) 5.03×10^{-58}m^{-3}, (b) 5.03×10^{-69} m^{-3}.]

2.3 Given that the sun's radius is 6.96×10^{8}m, use the answers of 2.1 and 2.2 to work out R, the radius of the universe for Olbers paradox to apply. Express R in light-years. [**Ans.** 4.15×10^{23} light-years].

2.4 Repeat 2.3 for galaxies, given that a galaxy can be considered very crudely as a sphere of radius 3×10^{4} light-years. [**Ans.** 2.49×10^{11} light-years].

PROBLEMS CHAPTER 3

3.1 With respect to the fixed stars Mars rotates about its axis in 24 hours 37 minutes and takes 687 terrestrial days to go once round the sun. How longer than the Martian sidereal day, to the nearest second, is the Martian solar day? [**Ans.** 133s].

3.2 In what year A.D. will the Gregorian calendar be wrong by a full day? [**Ans.** 4901 A.D.]

3.3 Assume that the day has been lengthening steadily at a rate of 1.6ms per century, and calculate the number of hours in a day 4.6×10^{9} years ago. [**Ans.** 3.6h].

3.4   Show that the angular momentum L of the Moon in its orbit around the Earth is given by
$$L = m(GMr)^{1/2}$$
where m = mass of Moon, M = mass of Earth, r = Earth-Moon distance, G = gravitational constant. (Assume the orbit is circular).

3.5   The angular momentum of the Earth is $I\Omega$, where I = moment of inertia and Ω = angular frequency of axial rotation. Assume loss of angular momentum by the Earth by tidal friction is equal to the gain of orbital angular momentum of the Moon. If the day has lengthened at a rate of 1.6ms per century, how far away was the Moon 3.5×10^9 years ago? ($I = 8.04 \times 10^{37}$kgm^2, $\Omega = 7.29 \times 10^{-5}$ radians per second at present time, $m = 7.35 \times 10^{22}$kg, $M = 5.98 \times 10^{24}$kg, $r = 3.84 \times 10^8$m at present time. Symbols as in Problem 3.4).   [**Ans.** 5.68×10^6m (about the earth's radius)]

PROBLEMS CHAPTER 4

4.1   Assume the earth is spherical with radius R and show that the acceleration of gravity g varies with latitude ϕ according to
$$g = g_p - \Omega^2 R \cos^2 \phi$$
where g_p is the acceleration at the poles.

4.2   Describe what happens to someone's weight when he walks, (a) east, (b) west, (c) north, (d) south.

4.3   The velocity of sound in air of density 1.29kgm^{-3} at $0°$C is 0.331kms^{-1}, in water it is about 1.5kms^{-1}, and in marble of density 2.6×10^3kgm^{-3} it is 3.8kms^{-1}. In what ratio to one another are the elastic restoring-force constants of air, water and marble?   [**Ans.** $1 : 1.6 \times 10^4 : 2.7 \times 10^5$]

4.4   Take the data in 4.3 and calculate the wavelength of sound at 10Hz in (a) air, (b) water, (c) marble.   [**Ans.** (a) 33m, (b) 150m, (c) 260m.]

4.5   If the earth's magnetic field-strength decreases at a rate of 0.05% per annum, how long will it take before the field is only half as strong as it is today?   [**Ans.** 1,386 years]

PROBLEMS CHAPTER 5

5.1   In the radioactive decay of uranium-238 (U^{238}) to lead-206 (Pb^{206}) eight α-particles (He^4) are produced per uranium atom. U^{238} is known to decay with a half-life of 4.51×10^9 years. If the decrease in U^{238} is neglected and the α-particles are retained as helium gas, how much helium is produced by 1g of U^{238} a year?   [**Ans.** 2.07×10^{-11}g] (Note that the fraction decaying per unit time is equal to $\log_e 2$ divided by the half-life).

5.2   A mineral sample is found to contain 5.6mg of U^{238} and 16 μg of helium gas. Assuming no helium has been lost and the decrease in U^{238} is negligible, calculate the age of the mineral.   [**Ans.** 138My)]

5.3 Depict the history of the earth on a 12 hour scale in which the origin of the earth 4,600My ago took place at midnight and the present time is the following noon. At what times, roughly, did (a) invertebrates appear, (b) the Alps, Himalayas and Rockies form and (c) man appear?

[**Ans.** (a) 10:26, (b) 11:49, (c) 11:59]

5.4 In an earthquake of magnitude 6.5 the period of the dominant tremor was 0.5s. What amplitude of ground motion was there?   [**Ans.** 1.6m]

5.5 The cycles for the precession of the equinoxes, the obliquity of the ecliptic, and the orbital shape are respectively 25,800, 40,000, and 92,000 years. How often does it occur that (within a time span of 4,000 years) the obliquity is maximum, the orbit is at most elliptical and the northern hemisphere experiences winter at aphelion?

[**Ans.** 280,000y corresponding to 11 precessional cycles, 7 obliquity cycles and 3 orbital cycles]

PROBLEMS CHAPTER 7

7.1 The electric energy of repulsion between two protons in a vacuum is given by the equation

$$E_e = \frac{e^2}{4\pi\epsilon_0 r}$$

where e is the elementary charge 1.602×10^{-19}C (coulombs), ϵ_0 is the permittivity of free space 8.854×10^{-12}Fm^{-1} (farads per metre), and r is the distance apart. The gravitational energy of attraction is given by

$$E_g = G\frac{m_p^2}{r}$$

where G is the gravitational constant 6.672×10^{-11}Nm^{2}kg^{-2} (newton-metre squared per square kilogram), and m_p is the mass of the proton 1.673×10^{-27}kg. Calculate the ratio E_g/E_e.   [**Ans.** 8.097×10^{-37}]

7.2 Calculate the electric energy of repulsion in electron-volts (eV) for two protons in a nucleus 3×10^{-15}m apart. (1eV = 1.60×10^{-19} joules).

[**Ans.** 0.480MeV]

7.3 From 7.2 calculate the electric energy of repulsion per particle E_p and obtain the ratio E_p/E_n, where E_n is the typical nuclear binding energy per particle, 8MeV. The result is a measure of the relative strengths of the electrostatic and strong nuclear interactions.   [**Ans.** 3×10^{-2}]

7.4 The time for the strong interaction to exert its influence in a nuclear collision is of the order of the time it takes for light to cross the nucleus. Show that this time is of order 10^{-23}s. On the other hand, decays of heavy particles into nucleons and mesons under the weak interaction take of the order of 10^{-10}s, and mesons themselves decay under the same interaction in about 10^{-8}s. What does that suggest about the relative strengths of weak and strong interactions?

7.5 In a nuclear fission process uranium-235, with a binding energy per nucleon of 7.6MeV, splits into nuclei in which the binding energy per nucleon is about 8.5MeV. How much energy is released per atom?
[**Ans.** 212MeV]

7.6 Given that the mass of a nucleon is 1.67×10^{-27}kg, what power in megawatts will be released by the fission of one kilogram of uranium-235 in one day? [**Ans.** 999MW]

7.7 In a nuclear fusion process deuterium, with a binding energy per nucleon of 1.11MeV, fuses to form helium-4, with a binding energy per nucleon of 7.07 MeV. How much energy is released per atom? [**Ans.** 11.9 MeV]

7.8 If one kilogram of deuterium is converted into helium a day, how much power would be generated? [**Ans.** 6,610MW]

7.9 What is the energy in electron-volts of (a) an infrared photon of wavelength 10.6 μm emitted by a CO_2 laser, (b) an infrared photon of wavelength 1 μm, (c) a blue photon of wavelength 0.4 μm and (d) an ultraviolet photon of 1000Å? (Velocity of light $= 2.998 \times 10^8$ms^{-1}, Planck's constant $h = 6.626 \times 10^{-34}$ JHz^{-1} $= 4.136 \times 10^{-15}$eV Hz^{-1}).
[**Ans.** (a) 0.117eV, (b) 1.240eV, (c) 3.10eV, (d) 12.4eV]

7.10 What intensity of radiation is emitted by a blackbody of 21°C? (Absolute zero of temperature $= -237.17$°C, Stefan's constant $= 5.670 \times 10^{-8}$ Wm^{-2}K^{-4}) [**Ans.** 424.6 Wm^{-2}]

7.11 In the previous problem, what was the wavelength at maximum intensity? (Wien's constant $= 2.898 \times 10^{-3}$m.K). [**Ans.** 9.851 μm]

7.12 If the sun emits like a spherical blackbody of radius 6.96×10^8m and temperature 5,800K (a) what is the total radiation power emitted, and (b) what is the wavelength at maximum intensity?
[**Ans.** (a) 3.91×10^{26}W, (b) 0.500 μm]

7.13 What is the thermal energy kT in electron-volts at (a) 21°C, (b) 5,800K, (c) 4K? (Boltzmann's constant k $= 1.381 \times 10^{-23}$JK^{-1} $= 0.8617 \times 10^{-4}$ eVK^{-1}). [**Ans.** (a) 0.0253eV, (b) 0.500eV, (c) 0.345meV]

7.14 A molecule has a binding energy E of 2eV. If the probability of dissociation is proportional to exp $(-E/kT)$, calculate the rise in temperature above 300K which increases the dissociation rate by a factor of 10.
[**Ans.** 9.2K]

7.15 What would be the radius of the earth (mass 5.98×10^{24}kg) if the escape velocity equalled the velocity of light, i.e. if the earth were a black hole?
[**Ans.** 8.88mm]

7.16 What is the average thermal velocity of a neutron at 300K? (Mass of neutrons $= 1.675 \times 10^{-27}$kg). [**Ans.** 2.724kms^{-1}]

7.17 In the atmosphere, 100 to 600 km above the earth, the temperature is about 1400K. What is the thermal velocity of a hydrogen molecule, mass 3.348×10^{-27}kg?
[**Ans.** 4.162kms^{-1}, i.e. about a third of the escape velocity]

7.18 Calculate the ratio of the Coriolis force to gravitational force of a man strolling at 5kmh^{-1} through Wivenhoe Park in Colchester, England (longitude $0°57'$ E, latitude $51°52'$ N).   [**Ans.** 1.62×10^{-5}]

7.19 In the situation of the previous problem what is the period of inertial oscillation?   [**Ans.** 15.2 hours]

7.20

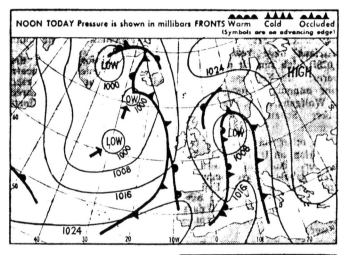

NOON TODAY Pressure is shown in millibars FRONTS Warm Cold Occluded (Symbols are on advancing edge)

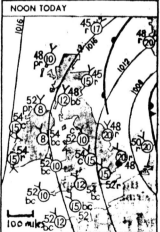

NOON TODAY

From this Meteorological Office weather map reprinted from *The Times* of May 6th, 1978, showing isobars in millibars and windspeeds in miles per hour in the circles, estimate the geostrophic wind-speed in miles per hour at Colchester (marked with a dot on the smaller map, and latitude and longitude given in Problem 7.18) and compare with the 20 miles per hour forecast. (Density of air = 1.29kgm^{-3}, 1 bar = 10^5Nm^{-2}, newtons per square metre).

[**Ans.** about 45 miles per hour]

b — blue sky; bc — half clouded; c — cloudy; o — overcast; f — fog; d — drizzle; h — hail; m — mist; r — rain; s — snow; tdr — thunderstorm; p — showers; prs — periodical rain with snow.

7.21 Supersaturated air containing 0.5% by weight of water vapour encounters dust overland which causes it to form a cloud of water droplets. If 50% of the vapour condenses in this way and the latent heat released all goes to raise the temperature of the air, what is the temperature rise? (Latent heat of condensation $= 2.5 \times 10^6 \text{Jkg}^{-1}$, specific heat of air $10^3 \text{J K}^{-1} \text{kg}^{-1}$).

[**Ans.** $6.3°C$]

PROBLEMS CHAPTER 8

8.1 At a certain altitude the scale height of the atmosphere is found to be 7.113km. What is the temperature? (Gas constant per mole $= 8.314 \text{Jmol}^{-1}$ K^{-1}, molecular weight of air $= 28.970$, $g = 9.807 \text{ms}^{-2}$). [**Ans.** $-30.1°C$]

8.2 Assume that the sun and planet X emit radiation like blackbodies and that the fraction of solar radiation incident on planet X which is absorbed is 0.65. Assume further that the atmosphere of planet X enhances the surface temperature through the greenhouse effect by a factor $2^{1/4}$. Between what limits must the orbit of planet X lie in order for water on the surface to remain liquid? (Temperature of the sun $= 5800K$, radius of sun 6.96×10^8m). [**Ans.** Between 9.59×10^{10}m and 1.79×10^{11}m]

8.3 Express the previous answer in AU (1 AU (astronomical unit) $= 1.496 \times 10^{11}$m, the mean earth-sun distance). [**Ans.** 0.64 to 1.20 AU]

8.4 If the solar system is regarded as a random distribution of ten planets in circular orbits in one plane whose radii may lie anywhere between 0 AU and 30 AU, calculate the probability that a planet X occurs with liquid water on its surface. [**Ans.** 0.19, or about one chance in five]

8.5 In the previous problem, replace ten planets with four (minor) planets all within 4 AU of the sun, and recalculate the probability.

[**Ans.** 0.56, somewhat better than even chance]

8.6 Radiation falls on a circular disc set at right-angles to the direction of propagation of the radiation. Show that if the disc is replaced by a sphere of the same radius, the average intensity over the illuminated surface of the sphere is just half what it was for the disc.

8.7 Show that if the intensity of solar radiation reaching the earth is 1.35kWm^{-2} and if 44% of this reaches the earth's surface, the intensity at the surface averaged over latitude, season and rotation is 149Wm^{-2}.

8.8 If the solar constant is 1.35kWm^{-2}, 35% of which is reflected, and the radius of the earth (considered spherical) is 6360km calculate the net input power to the whole earth. [**Ans.** 1.12×10^{11}MW (megawatts)]

8.9 In a thundercloud an updraught of 30kmh^{-1} allows charge to build up an electric field. Roughly what maximum field could be sustained? (Density $= 1.29 \text{kgm}^{-3}$, permittivity of air $= 8.9 \times 10^{-12} \text{Fm}^{-1}$).

[**Ans.** $3.2 \times 10^6 \text{Vm}^{-1}$]

8.10 Assume that the mobility of an ion is inversely proportional to its mass (which is usually the case). Calculate the percentage reduction of mobility caused by $(H_3O)^+$ picking up one molecule of water. **[Ans.** 49%]

8.11 A radioactive source of 5 millicuries emitting particles of energy 1.4MeV is accidentally left unscreened close to a person's body for 45 minutes. If one-tenth of all radiation emitted is absorbed by 2kg of the body what dosage in grays has occurred? (1 curie $= 3.7 \times 10^{10}$ disintegrations per second, $1eV = 1.602 \times 10^{-19}$ joules). **[Ans.** 5.6m Gy]

8.12 Calculate the extra dosage of radiation suffered by a mountaineer who climbs Mount Everest, assuming he spends three days at an average altitude of 8,000m. **[Ans.** 3.3 μGy]

8.13 Estimate the area of tropical desert on the earth's surface, and calculate what solar power could be derived from a total cover of 10% efficient solar cells.

[Ans. Approximate area $= 2.5 \times 10^{13} m^2$, average flux $= 200 Wm^{-2}$,
therefore total power $= 5 \times 10^8 MW$]

8.14 The natural abundance of deuterium is 0.015% of all hydrogen. Applying that figure to sea-water and taking the mass of water in the oceans to be 1.4×10^{21}kg, how many years would it take to convert all the deuterium in the oceans into helium, limiting the energy production (see Problem 7.8) to 10^9MW in order to avoid heat pollution. **[Ans.** 4.25×10^8 years]

Appendix

Table A1 — Symbols for Units

Quantity	Unit	Symbol
Time	Second	s
Space	Metre	m
Mass	Kilogram	kg
Force	Newton	N
Charge	Coulomb	C
Current	Ampère	A
Potential	Volt	V
Capacitance	Farad	F
Energy	Joule	J
Power	Watt	W
Frequency	Hertz	Hz
Temperature	Degree Kelvin	K
Radioactivity Dose	Gray	G

Table A2 — Symbols for Quantity

Amount	Prefix	Symbol
10^{-18}	Atto	a
10^{-15}	Femto	f
10^{-12}	Pico	p
10^{-9}	Nano	n
10^{-6}	Micro	μ
10^{-3}	Milli	m
10^{-2}	Centi	c
10^{3}	Kilo	k
10^{6}	Mega	M
10^{9}	Giga	G
10^{12}	Tera	T

Table A3 — Physical Constants

Quantity	Symbol	Value
Speed of light in vacuum	c	$2.998 \times 10^8 \mathrm{ms}^{-1}$
Permittivity of vacuum	ϵ_0	$8.854 \times 10^{-12} \mathrm{Fm}^{-1}$
Planck's constant	h	$6.626 \times 10^{-34} \mathrm{Js}^{-1}$
Elementary charge	e	$1.602 \times 10^{-19} \mathrm{C}$
Gravitational constant	G	$6.672 \times 10^{-11} \mathrm{Nm}^2\mathrm{kg}^{-2}$
Mass of electron at rest	m_e	$9.110 \times 10^{-31} \mathrm{kg}$
Mass of proton at rest	m_p	$1.673 \times 10^{-27} \mathrm{kg}$
Mass of neutron	m_n	$1.675 \times 10^{-27} \mathrm{kg}$
Boltzmann's constant	k	$1.381 \times 10^{-23} \mathrm{JK}^{-1}$
Stefan's constant	σ	$5.670 \times 10^{-8} \mathrm{Wm}^{-2}\mathrm{K}^{-4}$
Electron-volt	eV	$1.602 \times 10^{-19} \mathrm{J}$
Angstrom unit	Å	$10^{-10} \mathrm{m}$

Table A4 — Some Useful Quantities

Quantity	Symbol	Value
Astronomic unit	AU	$1.496 \times 10^{11} \mathrm{m}$
Parsec	pc	$3.086 \times 10^{16} \mathrm{m}$
Light-year	ly	$9.461 \times 10^{15} \mathrm{m}$
Solar mass	$M_\odot$	$1.989 \times 10^{30} \mathrm{kg}$
Solar radius	$R_\odot$	$6.960 \times 10^8 \mathrm{m}$
Solar luminosity	$L_\odot$	$3.90 \times 10^{26} \mathrm{W}$
Solar constant		$1.35 \mathrm{kWm}^{-2}$
Earth mass	$M_\oplus$	$5.977 \times 10^{24} \mathrm{kg}$
Equatorial radius, a	a	$6.378 \times 10^6 \mathrm{m}$
Polar radius, c	c	$6.357 \times 10^6 \mathrm{m}$
Mean radius $(a^2 c)^{1/3}$	$R_\oplus$	$6.371 \times 10^6 \mathrm{m}$
Surface area		$5.101 \times 10^{14} \mathrm{m}^2$
Land area		$1.48 \times 10^{14} \mathrm{m}^2$
Ocean area		$3.63 \times 10^{14} \mathrm{m}^2$
Mean land elevation		$825 \mathrm{m}$
Mean ocean depth		$3,770 \mathrm{m}$
Ocean mass		$1.42 \times 10^{21} \mathrm{kg}$
Mass of atmosphere		$5.30 \times 10^{18} \mathrm{kg}$
Equatorial surface gravity	g	$9.780 \mathrm{ms}^{-2}$
Moon mass	$M_{\mathbb{C}}$	$7.349 \times 10^{22} \mathrm{kg}$
Moon radius	$R_{\mathbb{C}}$	$1.738 \times 10^6 \mathrm{m}$
Moon distance from Earth		$3.844 \times 10^8 \mathrm{m}$

Index

NOTES

NOTES

NOTES

NOTES